OTHER PUBLISHED BOOKS BY THE AUTHOR

- Sundarban's Mangrove - Wildlife and Ecology (1985)
- Environmental & Resources of Tropical & Temperate Forests (1990)
- Environment and Herb Shrub Flora (1991)
- Himalayan Ecology and Environment (1992)
- Tree and the Environment (1993)
- Forest Plants of Eastern India (1992)
- Mine Environment and Management (1992)
- Plants, Wildlife and Man (1993)
- Sundarbans Mangrove - India (1995)
- Wetland Ecology (1996)
- Biodiversity endangered (2002)
- Megadiversity Conservation (2003)
- Plants of Jalpaiguri and Darjeeling (2000)
- Wildlife and Ground flora (2000)
- Golden Triangle of Eco-education Tourism (2004)
- Fish and the Scientist (2000)
- *Mach Niye* (in Bengali) (2000)
- Flora & Fauna of Andaman & Nicobar Islands (In Bengali) (2003)
- Project Tigeı Reserves (2004)
- Deforestation & Perilous land degradation – Foresters unique role in amelioration (2005)
- Sundarbans and Andaman Mangroves (under publication)
- Wetlands - A Vanishing Resource in India (under publication)

BIODIVERSITY OF MANGROVES

A.B. CHAUDHURI
Indian Forest Service (Retd.)
Ex-Director, Forest Survey of India

2007

DAYA PUBLISHING HOUSE

DELHI–110 035

ISBN 81-7035-436-6

Published by : **Daya Publishing House**
1123/74, Deva Ram Park
Tri Nagar, Delhi - 110 035
Phone: 27383999
Fax: (011) 23244987
e-mail : dayabooks@vsnl.com
website : www.dayabooks.com

Showroom : 4760-61/23, Ansari Road, Darya Ganj,
New Delhi - 110 002
Phone: 23245578, 23244987

Laser Typesetting : **APCON**
Delhi - 110 064

Printed at : **Chawla Offset Printers**
Delhi - 110 052

PRINTED IN INDIA

DEDICATION

This book is dedicated to
the generations of field foresters
of Sundarbans and Andaman and Nicobar Islands.

—**A.B. CHAUDHURI**

Acknowledgements

The author gratefully acknowledges the co-operation of various survey authorities (Botanical, Zoological and Anthropological) and of the Department of Forests for acceding the request to utilize some of their photographs and permitting him to use them.

He is happy to expresses his gratefulness to Dr. Nair of Centre of Earth Sciences for permission to reproduce some of his observations on Andaman and Nicobar Islands.

A.B. CHAUDHURI

Foreword

Chaudhuri's choice of World's two most resourceful and beautiful mangrove forests and presentation of multi-faceted qualities and resources proves his ingenuity and profound experience of the difficult terrain and various concepts attached to them. While he sings the song of the beautiful dreamlands, he also exposes the severity of hostile terrain infested with man eating tiger, crocodile, sharks and venomous snakes. Although these two sites fall outside the two megadiversity and hot spots zones (Western ghat and North eastern Indian forests), Chaudhuri considers these areas having more versatile in varieties of biota, play of innumerable physiographic features, endemism and biogeographical linkage (Andaman Islands) being more diverse. Chaudhuri authors many books including two on Sundarbans which are (i) Sundarbans mangrove (Ecology and Wildlife, 1989) and (ii) Mangroves of the Sundarbans (IUCN Publication 1995). About Andaman his Resources Survey reports as Director of Forest Survey of India and expert note on 'Protected Areas' speak of biotal richness etc. of the Bay Islands. These books and reports testify to Chaudhuri's, original work.

Most outstanding work centres round exposure of the behaviour pattern of man eating tigers which kills more than 40 people a year. Selection of 21 true stories about Man-Tiger interaction is unique which also reveal how the people in spite of high degree of natality want to coexist with the tiger.

In my official capacity as the Principal Director I had opportunities to visit mangrove and adjoining areas and interact with people and I find Chaudhuri's observations speak the mind of the forest's fringe people. The people whom I met did not ever speak against the man-eaters categorically except that they deserved handsome compensation money for the affected families.

Sundarbans cover a difficult terrain and needs development in various fields.

His selection of protected areas and recording their biota (in Andaman and Nicobar Islands) seems to be very creditable. He presents site-specific biotal check lists. Readers will find this book very handy in having a quick glimpse of various resources problems and conservation strategy outlined by Chaudhuri of the two sites. He also gives some details of the role of marine water which joins both the sites. In an earlier publication [*Golden Triangle of Eco-education Tourism* (2004)] Chaudhuri has presented various qualities of these forests which should be very useful for the eco-education tourists. The readers may refer to that publication.

Chaudhuri has expressed concern about the heavy rehabilitation of people on the Island, felling of forests to create agricultural land, extension of Industries which would be ecologically and environmentally disastrous for the Islands. The fragilities of expressing tropical forest soil has not been understood by the administrators it seems.

Dr. T.K. ROY
Retired Adviser & ex-officio
Principal Director
Animal Resources Development Department
Government of West Bengal,
Kolkata – 700 001

Preface

Of the total forest cover of 637,293 sq. km., the Mangrove forests cover are only 4,871 sq. km or about 0.15 per cent.

Nevertheless these special vegetation type play a very distinctive role in the biotal ecology and coastal management. Moreover, these forests have a rich assemblage of planktonic flora, a varieties of invertebrate and some vertebrate fauna; such areas have also distinctive ornithological and Ichthyological (fish) fauna having innumerable behavioural patterns.

The scenario of forest cover in India is not very hopeful. Of 3.2 million sq. km of Geographical area, India has 0.637 sq. km under forest (F.S.I. 1999 report) which form about 19.39 per cent of the G.A. of the Country. The following chart may give a clear view.

India has largest area in Sundarbans mangrove followed by Andaman and Nicobar Islands. Together they form an area of 5,15,800 ha. It must not be forgotten that the combined area of Sundarbans of India and Bangladesh form the largest single patch of mangrove area in the World.

For interest of the readers a chart indicating floral and fauna diversity of India is produced although the mangroves have special biota to offer.

The chart below represents an estimation of plant resources of India. Some species-rich areas have been mentioned in several other charts. The areas under discussion which are Estuarine Sundarbans mangrove and Bay Island Mangroves are equally rich in estuarine and marine biota which have many endangered and endemic species. As the richness and behavioural patterns of hundred of vertebrate and invertebrate fauna cannot be discussed due to brevity of space, only a few have been brought to the sphere of discussion.

Along with the interesting flora and fauna of the two mangrove areas chosen in this study (The Sundarbans mangrove and Andaman-Nicobar mangroves) it should be relevant to indicate some rich floral and faunal zones in India which are presented in the chart. However, the author feels that there are few areas as facinating and resourceful as the mangroves of Sundarbans and Andamans.

Class	Area in sq. km.	Per cent of Geographical Area
Dense forest	377,358	11.48
Open Forest	255,064	7.76
Mangrove Forest	4,871	0.15
Sub Total	6,37,293	19.39
Scrub Forest	51,896	1.58
Non Forest	2,598,074	79.03
Total	3,287,263	100.00

[Source: F.S.I.]

However, India needs tree cover all over the country in view of scattered growing human and cattle population, expansion of industries, mines and large scale denudation. The conservation effort of 28 tiger projects will likely to save bulk of India's vegetation types and various species of wild animals (except the high hills). The mangroves of the study area have immense varieties of biota which have also the following qualities:

- They bring joy to the visitors, protect land mass from the impact of cyclone, their roots bend, soil and the estuary serves as nursery ground for various fauna.
- Various living organisms act in a rhythmic sequence.
- They have magical qualities of contrast, light and shade, peace and conflicts, facts and fancies.

FLORA/FAUNAL BIODIVERSITY OF INDIA

(*Species estimation approximate*)

INDEX OF MINIMUM SPECIES

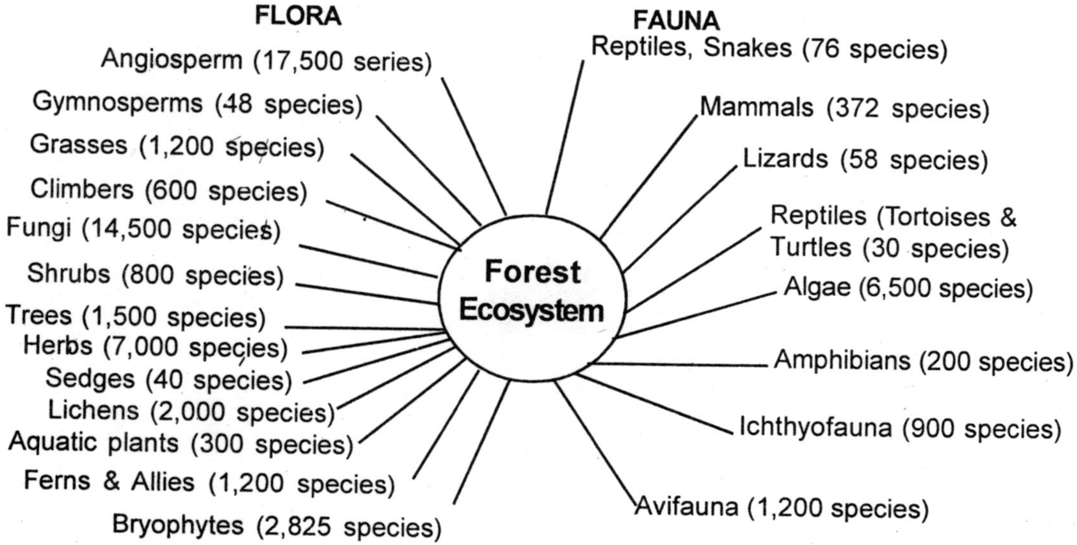

Chart 1

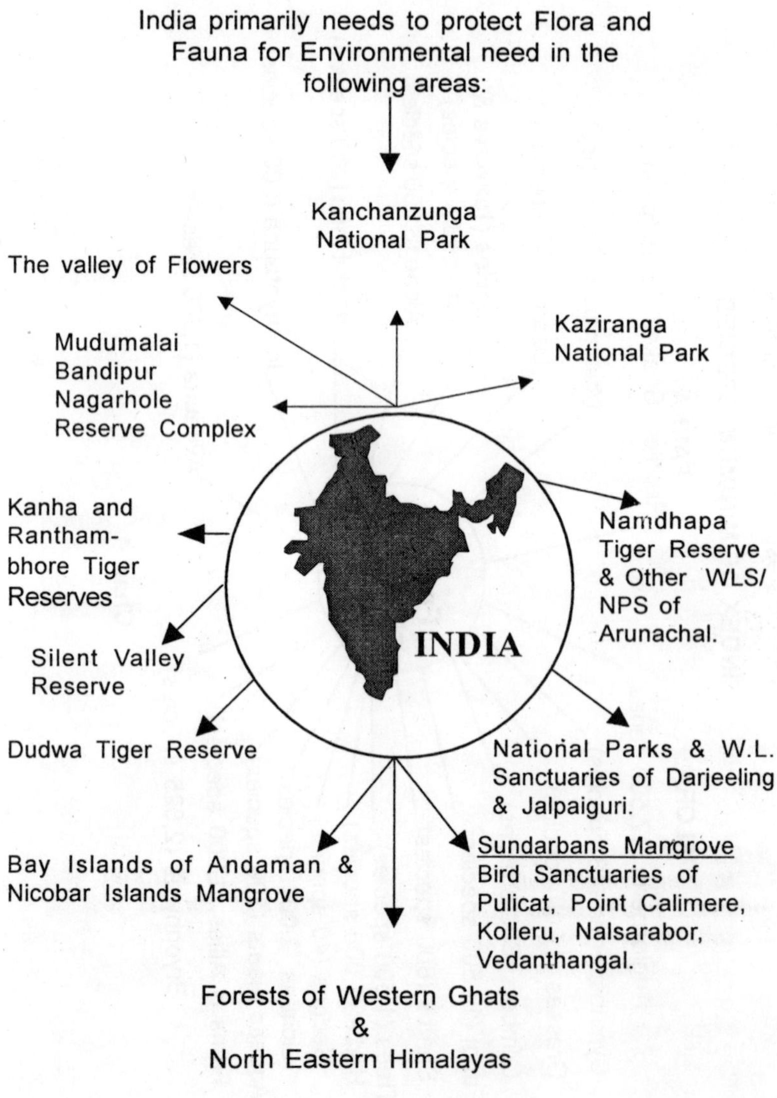
India primarily needs to protect Flora and Fauna for Environmental need in the following areas:
Kanchanzunga National Park
The valley of Flowers
Kaziranga National Park
Mudumalai Bandipur Nagarhole Reserve Complex
Kanha and Ranthambhore Tiger Reserves
Namdhapa Tiger Reserve & Other WLS/ NPS of Arunachal.
INDIA
Silent Valley Reserve
Dudwa Tiger Reserve
National Parks & W.L. Sanctuaries of Darjeeling & Jalpaiguri.
Sundarbans Mangrove
Bird Sanctuaries of Pulicat, Point Calimere, Kolleru, Nalsarabor, Vedanthangal.
Bay Islands of Andaman & Nicobar Islands Mangrove
Forests of Western Ghats & North Eastern Himalayas

Chart 2

- Mangrove soil and water throb with life processes round the clock.
- Mangroves display intricate balance in nature.
- In mangrove one may breathe in rhythm, sing in rhythm, talk in rhythm, think in rhythm and dance in rhythm.
- Mangroves are 'Wild, 'Wonderful' and the best for better than the best in India such as - Wild Ass in the Rann of Kachch, Gahirmatha marine Sanctuary of Sea-turtle in Orissa, Painted Stork and Flamingos of Wetlands, Kashmir Slag of J & K, Tigers of Ranthambore, Phayre's spectacled monkey of Tripura and many other such.

There are many more qualities which have been enumerated in various chapters.

Author's close study and findings on the flora and fauna of the two areas has been recorded in some of his publications (refer Bibliography).

Most outstanding work done by him in Sundarbans was from 1970-1973 on the behavioural pattern of man-eating tigers and breeding biology and ecology of some birds, honey bee, fish etc. These have been supplemented by visits from time to time in the areas. His involvement with the biota of Bay Islands was from 1978 to 1983 and there after in 1984, 1985 and 1989 (Bibliography).

As such, most of the data, maps and charts produced in this publication are also from author's work mostly recorded in two books which are-

- Sundarbans Mangrove (Ecology and Wildlife, 1989)
- Mangroves of Sundarbans: Vol. One India (IUCN – The World Conservation 1994)

Various data have been updated.

The report of Dr. S.C. Nair, Centre of Earth Sciences, Trivandrum on 'Natural Resources Conservation and Development' in the Bay Islands has been adequately utilized.

The reader should peruse the maps and charts to appreciate the subject matter discussed in the book. It is expected that the readers will get scores of facts and figures on the ecological and biological importance of the two areas. Various factors affecting the very existence of the Bay Islands have been discussed and conservation strategy outlined.

The salient features of the two sites have been discussed separately in two parts, Part I – dealing with Sundarbans and Part II with Andamans and Nicobar Islands.

A.B. CHAUDHURI

131, NSC Bose Road
Block 10, Flat 4
Kolkata - 700 040
Ph. : 24716734

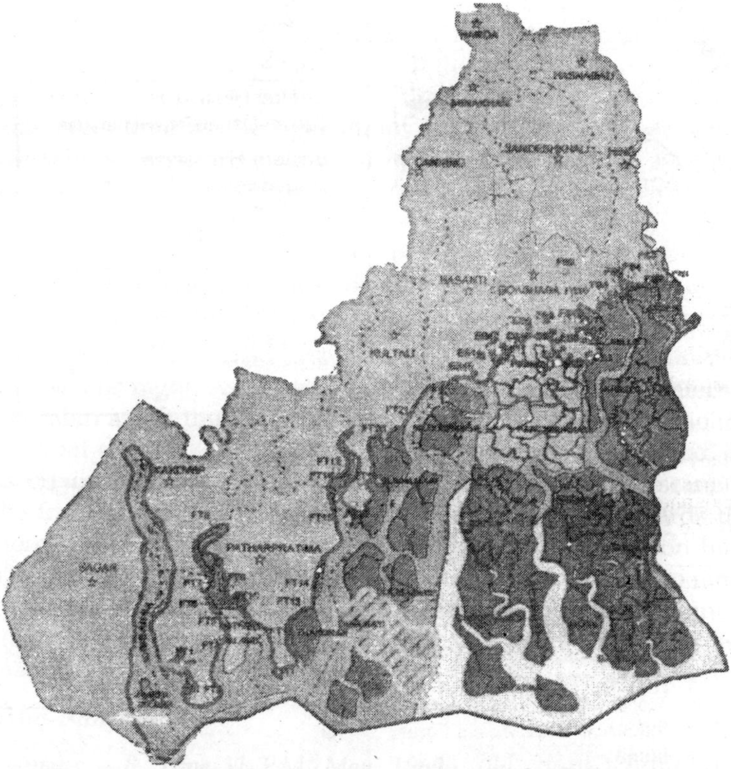

SUNDARBANS MANGROVES FLORA, FAUNA, MAN-ANIMAL INTERACTIONS AND CONSERVATION STRATEGY PART 1

GOLDEN TRIANGLE OF ECOEDUCATION

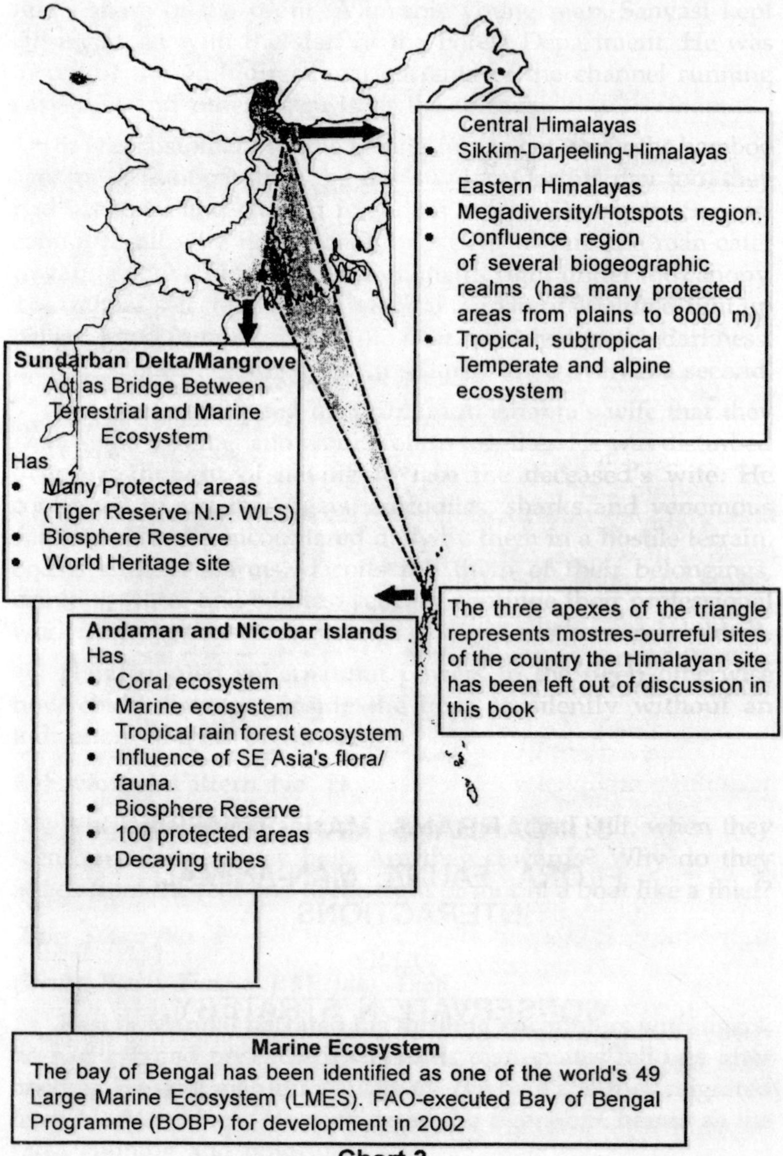

Chart 3

Contents

PART II

FLORA AND FAUNA OF ANDAMAN AND NICOBAR MANGROVES

(RESOURCES, CONSERVATION STRATEGY, PROBLEMS, REMEDY)

Introduction

PART – I

MANGROVE FORESTS OF SUNDARBANS ESTUARY - A HABITAT OF BEAUTY AND HORROR AND A FLORAL AND FAUNAL RESOURCE BONANZA

Whatever may be the derivations of the three appellations -

(i) Forests of Sundri (Heritiera minor),

(ii) Beautiful (Sundar) Forests and Animals,

(iii) Forests of ocean (Samudraban) —

the Sundarbans stand supreme in habitat maps of the world. Few areas exist in the world with such diversities in animal species that live in a complex physical, chemical, mechanical and animate environment. The plants and animals are in perfect balance in a rich biomass by their stresses and strains, thrusts and counter thrusts. Sundarbans offer a perfect example where different groups of living organisms live in a hostile and complex biosphere (biogeocoenosis) developing various adaptations (Osmo-regulatory mechanism, Endeogenous Self Sustaining Oscillation, Pneumatophores etc.).

This is a peculiar terrain where man-eating tigers live in the midst of man-eating ferocious crocodiles and with sharks, lizards, birds, porpoises, fish, turtles, snakes, honey bee, crabs, snails and innumerable vertebrate and invertebrate species. Even people working in the hostile estuarine tract develop peculiar habits, and the influence of myths, legends and the fisher-folk of the Sundarbans make the habitat as one of the most difficult but interesting for any biological study.

Mangrove forests until recently did not receive adequate attention despite their potentialities and being nature's richest laboratory and most complex ecosystem. Foresters were pioneer in bringing mangrove forests under scientific management and have collected adequate data on various resources. Though most of the mangrove forests in general are rich in faunal resources, they do not compare Sundarbans and Andamans in richness of fauna, their diversity and complexities which are result of special deltaic formation, geographic position and of interactions of mechanical, chemical, biological and physical environments. These complex environments have caused physiological and sometimes psychological changes in the entire animal population. No indepth study on behavioural pattern, breeding biology, growth pattern etc. on such animals has so long been made.

The author has worked for long years on Sundarbans of West Bengal and have made indepth extensive studies on behavioural pattern, breeding biology, growth, population etc. of tiger, honey bees, birds, crocodiles, crabs, fishes etc. Several such observation were earlier published in various journals as news letters and in symposia, workshops and conference literature. Vast, statistical, data and research findings which have been presented in this work are not generally available in the existing literature on wildlife. Plant animal relationship has also been closely studied and presented.

The author used the shape and size of pug marks of tigers in determining tiger population as early as 1970 and had ascertained the number of tiger between 112 and 120. Many such data and information have been incorporated in the work which are findings of indepth studies. These informations have made the work interesting and absorbing.

Much work has been done, but vast field of research still awaits to be unveiled.

All these make the entire terrain, both on land and water, impenetrably difficult to traverse by men or animals. It becomes immensely difficult for the predators and prey also to move easily inside the forests.

The vast expanse of swamps, studded closely with a network of tiny islands of mud flats support a low wooded forest of high density of virgin beauty. The rivers around tiny mud flats, form fantastic labyrinths of bifurcations and interlaces where pneumatophores give a trying time to the mysterious flamboyant man-eaters who have also to survive from venomous snakes, sharp toothed shark and lusty crocodiles. Such a hostile country is bound with natures masterpiece of creation the Tigers of Sundarbans. The resourcefulness, intelligence, agility as personified in such tigers have turned this magical species into a legendary myth. Superstition, and supernatural qualities attributed to the man-eaters, however fictitious they may appear, have their basis in facts. Locating its human prey, silently stalking him in ever closing circles for hours, high degree of intelligence to baffle trap gunner and trappers, diabolical understanding of the human mind, high precision in attacking human beings and crossing rivers, great power in controlling reflexes in mounting a boat, with least rocking, have made it a mythical animal. Blessings are invoked from Dakshin Ray (Tiger God), Narayani Ma, Banbibi, Kalu Khan, Sa Jungli, Gazi Sahib prior to entering the forests as it is thought that the animal have supernatural qualities of destruction of human beings who are helpless against its machinations.

The author has presented a series of true stories collected from a vast number of interrogaıory data and have depicted the pattern of behaviour and other biological features of the tigers of Sundarbans. The stories narrate exciting encounters with the tigers in the sinister beauty of the wild, swift flowing water, armoured deadlines of the crocodiles gliding noiselessly through water in quest of prey, mating call of the tigress and answering roar of the mate prowling silently along deserted jungle paths. Pathetic stories of honey collectors, fishermen and wood cutters give a grim picture of their struggle for survival against a powerful and cunning beast which is still considered mysterious. Yet these men have observed meticulous silence about their ruthless extermination; rather they wanted to coexist.

Tigers in the Sundarbans have taken a toll of about one thousand human lives in the last forty years. These are all reported cases; it is difficult to estimate the number of unreported cases. The victims are almost invariably poor villagers form the adjacent areas which are among the most backward in the country. Sheer economic necessity compels them to brave the hazards of the terrain over and over again. Fisherman, honey-collectors, wood-cutters and others engaged in work of an allied nature constitute the bulk of "permit-holders". These are the people for whom the Sundarbans tiger is a living entity- almost a part of their lives.

When such people talk of their experiences - no matter how tragic these are - the references to the tiger are rarely, if at all, in inimical terms. The tiger is "accepted". It is almost as if there is an elaborate ritual of ground rules to ensure coexistence. A priest accompanies them to the forest and invokes the blessings of Dakshin Ray - the principal folk deity and lord of the forest, in the robe of a warrior, armed with sword, gun, bow and arrows, and seated on a tiger – and also of Narayani Ma, Banbibi, Kalu Khan, Sa Jungli and Gazi Saheb to obtain immunity from any attack by man-eaters, crocodiles, Manasa (Snake) and other creatures.

The stories presented in this paper are from eyewitnesses and an attempt has been made to retain the original flavour. Whatever be the scientist's view on the overlay of superstition, the stories reveal certain interesting insights into the preying habits of the Sundarbans tiger.

A BRIEF PAST HISTORY

To trace back the history it may be recalled that the lands of 24-Parganas ceded to the East India Company by a treaty of 1737 which subsequently became the Jaigir of Lord Clive. The earlier and recent Sundarbans was made in 1770 by Mr. Claude Russel, Collector General 24-Parganas. Nothing much is known about the Sundarbans during the period intervening between the regime of Nawab and acquisition of Dewani by the East India Company. The whole of the District is a pure delta formed by the action of the river Ganga which brings down fine silt and sand from the Himalayas through the two great channels the Bhagirathi and the Padma. While the Bhagirathi is the original southward course of the Ganga, the Padma is a later east flowing branch of it. Some time during the 16th century the main volume of the Ganges passed from the Bhagirathi to the Padma. As a result land building process took place at a much faster rate in the delta region of the Bhagirathi on the west than in the delta region of the Padma in the east. This formation of new alluvial land naturally attracted the early settlers. This has practically started a progressive and continuous wave of reclamation of the forest areas. Records about the Sundarbans (including the portion in Bangladesh) show that the total area of the Sundarbans forests was taken as about 20,000 square kilometers by Clarke in 1895, 17, 500 square kilometers by Prain in 1903 and only 10,000 square kilometers by recent estimate of which hardly 4,000 square kilometers are included in West Bengal. Although by about 1879 a total area of about 4690 square kilometers of forests was declared as protected forests in 24-Paraganas District, leasing of new areas which included some of the notified protected continued. This was followed by deforestation and thence reclamation. As a result in 24-Parganas District alone during the last 90 years or so an area of about 750 square kilometers of forest land had been released for agricultural purpose. The whole area in 24-Parganas District was declared first as protected forests under a notification dated 7th December, 1878 and the boundaries of the remaining protected forests were fixed by the notification dated 19th April, 1926. The protected forests in the Basirhat Sub-Division of District (Basirhat Range present Tiger Reserve area) were constituted in the Reserve Forests as per Govt. notification dated 9th August 1928.

Sundarbans Forests in the Ganga-Brahmaputra delta, falling in India and Bangladesh are contiguous and comprise single patch of estuarine forest; it is the largest single patch of estuarine forest in the world, though cut up by countless rivers, creeks and waterways.

Besides the speciality of Sundarbans Mangrove, it has the maximum area in India; combined with the forests of Bangladesh this mangrove patch form the largest single mangrove patch of the world. The areas are given in the table below:

As such, Sundarbans and Andaman Forests taken together form a substantial percentage of estuarine forests.

The Indian Ocean and Western Pacific regions have 65 out of total of 90 typical mangrove species. Pacific America and Atlantic America both have 16 species each, while West America has 11 species. In India, the Gujarat coast has fewer species than the West coast and the both have combined fewer than the east coast species. Sundarbans, however, has the maximum number of species represented. Salinity in water and soil in Bangladesh is far less than in the West Bengal sector and these factors contribute to the emergence of a large number of species.

India	Area (ha)	Abroad	Area (ha)
West Bengal	4,18,888	Malaya	1,55,500
Sundarbans			
Andhra Pradesh		Puerto Rico	
Godavari delta	13,804	Fiji	2,000
Krishna delta	5,120	Kenya	32,500
Tamil Nadu			
Cauvery delta	2,640	Tanzania	10,500
Orissa			
Mahanadi delta	12,000	Zaire	20,000
Maharashtra			
West Coast	62,208	Indonesia	1,00,000
Saurashtra & Kutch	52,616	Mekong delta	2,50,000
Andaman & Nicobar Islands	1,15,200	Sundarbans in Bangladesh	12,97,500
	6,81,976		

ABOUT THE MANGROVES

The estuarine forests are 'Halophytes' that grow on swampy clay, muddy and saline substratum. Such forests in the Sundarbans have, generated a fascination on the human mind. Somc of the high fiats were once regarded as hiding places of Bengal Kings viz. Raja Basant Roy and his nephew Pratapaditya during the reign of Mughal emperor Akbar. After the fall of Bengal Kings anarchical conditions followed and the oppressions of Portuguese pirates put an end tot the prosperity to the tract. Ruins of old buildings and camping sites are still to be found in Netidhopani, Gosaba, Khatuajhuri and some other blocks. The dense jungles, man-eating tigers, sinister crocodiles, dampness of the swamps and nature's virulence offered protection to the Sundarbans forest. But human ingenuity outwitted the defence of nature. Vast areas were reclaimed for settlement. Illicit felling of trees by a large but poor population denuded the forests.

Botanical exploration dates back to 1776 by Roxbourgh who detailed the account of plants collected through Dr. Buchanan Hamilton and Dr. William Carey in his publication 'Hortus Bengalensis' published in 1814. C.B. Clarke in 1895 gave an account of the Sundarbans. Mr. Heining, a Forest Officer, made an effort to collect plants and also prepared a Working Plan. On his and Clarke's collection and on further work, David Prain published a flora of Bengal plants in 1903.

Ecological succession in the eastern Sundarbans is different from that in the Western Sundarbans. Sundarbans soils are alluvial while the soils of the deltaic region of South India are derived from trap and siliceous rocks. Rehabilitating the saline blanks existing in the estuarine forest at the moment has posed a major problem.

No mud flats like Ganga-Brahmaputra occur in the Andamans. There is no river in Andaman where delta can be formed. Since the islands are tips of submerged hills, the shores are steep. Coastal flat land does not occur. Evergreen semi-evergreen and littoral forests extend right upto sea face, leaving long strips of estuarine vegetation with hardly any appreciable width. Where the sea enters inland and forms creeks, extensive long stretches (with comparatively longer width) of Mangrove Swamps are found.

Forests in Sundarbans grow in various associations (associes) in different seral stages. As many as ten to twelve such distinct associations (associes) can be identified. In such a terrain lives euryphagous and eurythermal amphibious tigers which are man-eaters. They are even now considered by the local people to possess supernatural power. About 40 precious human lives are lost every year. The close associations of tiger, crocodiles, turtles, lizards, fishes, crabs, snails, snakes, barnacles, oysters, shrimps, honey bees and myriads of others animals with the plant life make a thorough study of the plants extremely hazardous and risky, but excitingly complex and interesting. As such, thorough study is yet to be undertaken on various aspects of estuarine plants, especially on their physiology, anatomy and ecology.

The present attempt does not cover details to –

(i) Classify the plants into various formations.

(ii) Prepare a list of plants which are purely estuarine and littoral.

(iii) Prepare field keys of common species for use by the foresters.

(iv) Give broad descriptions of typical estuarine species.

(v) Indicate a broad outline of plant succession.

The list is not exhaustive, although most of the important estuarine species and some on recently reclaimed land have been recorded. It must be borne in mid that the Indian part of Sundarbans have almost all typical estuarine species found in Bangladesh but fewer of littoral and sandy beach species.

This book has been designed to present a vast subject in a nut shell, due to brevity of space, to all readers of varied interests, the beauty, resources and horror of Sundarbans which are so close to vast population of Kolkata, yet far away due to hostile nature of the terrain. The special qualities of vegetation and myriads of invertebrate animals are of immense source of attraction to all nature lovers, educationists and explorers; but most unique episode centers round the interactions between Tiger and Man which is a special feature of this Mangrove. So the tiger episodes of the Mangroves of Sundarbans will remain the Centre of attraction to millions of people of this World for years to come.

PART – II

ABOUT THE BAY ISLANDS: ANDAMAN AND NICOBAR - AN INNOVATION

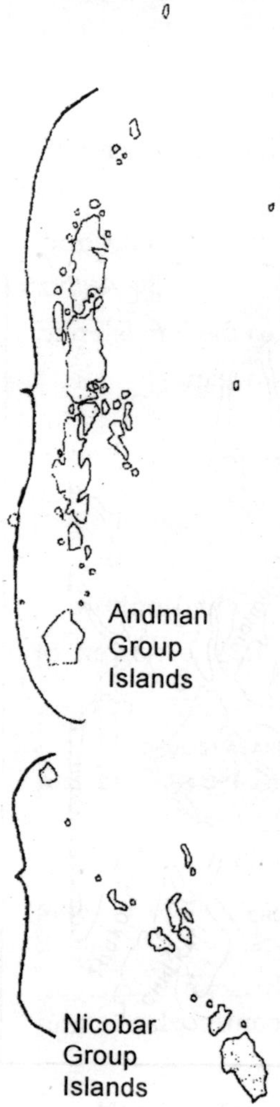

My beloved dreamland Andamans -
Who blasfemes you as 'Kalapani' (Black water);
You are my beloved dreamland
Sun's golden rays give you a fragrant bath.
Amidst an evergreen backdrop -
Mighty ocean hides many of your evergreen peaks -
Horizon unevils your lush-green endless valleys -
Underneath your green carpeted hill slops.
Are born scores of meandering streames.
My dreamland Andamans -
You have long stretches of beautiful sandy, clayey and loamy sea coasts.
That drowses under the shade of coconut palm,
And protected by sylvan evergreen mangrove forests.
Islands of dazzling – pearls-Andamans,
You dazzle in the bosom of blue ocean's ripples -
Your apparel is sylvan green -
Your motherly cosy lap shelters a few dacaying fossilised tribals

Who are silent sentinels of millions of years of any evolutionary processes?

Your tribals and the political prisoners have faced the onslaught of the unkind rulers.

The huge Octopus-shaped cellular Jail that encaged the freedom fighters

Still stands high to testify the stories of their bravery and ruler's repression.

My dreamland Andamans -

You are no longer 'Kalapani' of the past

You are heavenly, and an emblem of beauty and serenity.

I adore you with every fibre of my being

I pray 'ecoeducation' Tourist's will adore you in perpetuity

-By Author

The following epithets have been applied to the Bay Islands:

- A floating piece of pearl in emerald ocean.
- A floating golden Island in blue ocean
- An abode of million years old tribals.
- An enchanted fairy land.
- An unparallel beautiful island group
- A mini India that represents India's unified culture.
- A congregation of colourful forests in the bosom of emerald ocean.
- A nature's aquarium
- An unparallel assemblage of resource base of flora and fauna at a very small space.
- An unique assemblage of floral and faunal diversity.
- An exquisitely interesting assemblage of floating biota in an emerald Ocean's diverse species.

Andaman is an idyllic destination, earth clean blue seas, colourful coral reefs, golden beaches, coconut plams and aboriginal peoples.

Mangrove Forests of Andaman and Nicobar Island - A habitat of wide range of flora and fauna

The Islands remained in isolation and therefore undisturbed for millions of years. With the start of Scientific Forestry in late 1890s and early 1990, the forests were subjected to massive modification and they suffered drastic changes since the second World War and thereafter. Prior to the forester's became conscious of the evil effect of such changes, late Prime Minister Smt. Indira Gandhi said "These programmes (rehabilitation and commercial plantations) need clearance of substantial of virgin forest. When we drew up the plans, we were not fully aware that removing the forest cover may be so damaging as to threaten to reduce the islands to a waste land. Although review of the effect of deforestation on these islands, there would be little point in going to the trouble and expense of clearing forests if we are not assured of a long term favourable outcome".

The rehabilitation of more than 30,000 refugees of East Bengal, expansion of industries, extension of agricultural land by felling of virgin forests, conversion of forests over vast areas brought large scale deterioration of habitat. Even Mc Vean, IUCN Expert's cautious note was ignored till late 1980s.

The forests of Andaman and Nicobar islands are a national wealth and rightful heritage of mankind to be treated as such, to be perpetuated and utilized as human knowledge develops enabling him to harness the limited global living resources for a sustainable and egalitarian betterment of man.

There are not many more tropical islands in the world with as much of natural conditions remaining as in Andaman and Nicobar; hence it is our privilege to safeguard it. Worldwide scientific opinion is coming to accept this non-renewable nature of these forests and related ecosystems.

Had these islands only a marginal population, perhaps the deleterious consequences of unsuitable forestry practices could have been less harmful but with almost two lakhs of agriculturists resettled in this archipelago, forestry has to guarantee the primary agricultural requirements namely soil protection, water availability and maintenance of suitable agroclimate. In addition the dependence of the population for forest produce is greater here than in the mainland. Thatching, fodder, fuel, cane, bamboo, wood etc. are required in considerable quantities by the population and the Government is bound to supply to each resettled person a considerable quantum of each of these. Monoculture plantations or even the "naturally regenerated" more homogeneous forests cannot provide any of these.

Because of geological and topographical peculiarities, Andaman soil shows a marked variability from place to place along with other aspects such as slope, drainage etc. As a result vegetation show extreme variability. Extensive homogenous areas of any vegetation type cannot be located anywhere excepting the mangrove vegetation. Plantation species requires homogenous conditions and each species require a specific habitat. This intrinsic limitation to extensive plantation raising have contributed to past failures and many instances of unsatisfactory regeneration of worked areas. But during felling, coup demarcation or locating plantation areas, this ground variability is given little consideration.

This mosaic type of plant communities and the archipelago nature results in very localized distribution of species in Andaman and Nicobar islands. When potential extermination of species would also be high. Due to lack of detailed species inventories from most areas of these islands, even the loss is unrecorded.

All the evolutionary and biogeographic factors resulting in an extremely rich but fragile flora also bring about a unique but highly vulnerable fauna in oceanic islands. Being separated from the mainland by hundreds of kilometers of deep sea, the archipelago would not have any indigenous large mammal or biologically recently evolved animals which cannot withstand a sea journey. Avifauna and bats show considerable diversity, so also primitive litter or forest invertebrates that reached these islands before their links with the mainlands were cut off. Because fresh water is not plentiful, fresh water fishes and amphibians would be poorly represented. But marine coastal fauna and coral ecosystems would be extraordinary rich. Through the long years of isolation, the original animal inhabitants of the islands have evolved and specialized to such an extent that most of them are highly vulnerable to any habitat modification. So far about 55 species of mammals and 255 species of birds have been recorded in these islands. Excepting the avifauna, all other groups of higher terrestrial vertebrates i.e. fresh water fishes, amphibia, reptile and mammals are poorly represented. The avifauna is spectacular with a number of endemic endangered species such as the Andaman teal (*Anas gibberiforms albougularis*). Narcondum hornbill (*Rhyticeros undulates narcondami*). Nicobar megapode (*Megapodius freycient*), Nicobar pigeon (*Calaenas nicobarica nicobarica*) etc. There is a large influx of migrant coastal and shallow water sea birds and the list of endemic subspecies and races in quite long. Endemic mammals include the Nicobar crab eating macaque (*Macaca fascicularis*), the local subspecies of the Malay tree shrew (*Tupalia gils*), the palms civet (*Paradxurux tylleri*), the Andaman wild pig (*Sus andamanensis*) and a few species of rodents and bats. Dugong (*Dugong dugong*) though not endemic may be occurring in viable numbers only in Andaman and Nicobar islands within the Indian territory.

ANDAMAN AND NICOBAR TRIBES TODAY

The aboriginal tribals of Andaman and Nicobar islands and their future forms an important component of any comprehensive land use policy framed for these islands. Here again the absence of a clear cut policy based on reality has led to considerable confusion. The past history of the native inhabitants of the Andaman group of islands is not very clear. Evidently, the number of distinct negrito tribal groups had chosen a path of conflict with the 'foreign invaders' and had lost. Various tribal groups, originally spread over the North, Middle and South Andaman islands along with adjacent islands such as Ritchie's Archipelago and Rutland have disintegrated since the second half of last century and the remnants of half a dozen distinct tribes numbering less than twenty have been translocated to the straight island (603 hectare). Houses modeled on our concept of low-income Government servants' quarters have been constructed for them, with free electricity, medical facilities and doles handed out by the Tribal Development Authorities. 28 'racially impure' individuals now broadly called the Great Andamanese are looked after by the Administration.

From the Anthropologist's point of view, the Andaman and Nicobar islands have a unique position in India since this small area within its boundaries has 4 very 'primitive' tribes who have been insulated against extraneous influences. Their cultural and other adaptations and their harmony with the habitat have much to cared for.

Special physiographic features and other characteristics of the two sites are presented in 18 charts.

CONSERVATION OF MANGROVE FORESTS FOR HUMAN WELFARE

Coastal woodlands, known also as mangrove forests (or as littoral, estuarine or tidal forests) are nature's masterpieces of creation and ecological bonanza. These are, according to a UNESCO expert "Unique resources that are of great importance to the majority of people of this planet."

The inhaustible resources of the forests did not receive the attention either of the foresters or the administrators. During the last decade only the scientists assembled in various symposia and workshops and discussed about various resources, potential and problems of these forests. In 1978 a working group on mangrove (term used by all international organizations) was created at UNESCO's initiative. This working group in its first meeting resolved to undertake a biosphere inventory of mangrove lands and to determine their current status. In 1980 the Commission of Ecology (COE) of International Union for Conservation of Nature and Natural Resources (IUCN) created a working group on mangrove ecosystems to collate existing information. The Department of Science and Technology, Government of India formed a task force to evaluate the mangrove forest potentials of the country; later, the Department of Environment, Forest and Wildlife has been involved in conservation aspects of Mangrove.

Special features of mangrove forests that set them apart from other types of forests are:

- the trees and bushes grow below the high tide level,
- the root system is inundated by saline water at regular intervals,
- the forests are favoured by tropical humid climate accompanied by silt-laden rivers that form mud flats,

although some grow on extremely arid coats and on coral and rock islands.

- the trees develop special adaptation mechanisms such as breathing roots of various types, fleshy and thick leaves, salt glands, viviparous germination, sunken stomata, aqueous tissue etc.

Population pressure combined with major and multiple demands on the coastal resources for rehabilitation, urbanization, creation of ports, recreation facilities aquaculture has caused substantial damage to such forests locations of mangrove forests.

Unlike various types of tropical and temperate forests, mangrove forest are located in small areas covering about 84000 sq. km. along the coasts in most of the tropical maritime countries. The largest single patch of mangrove forests is in the Sundarbans of Bangladesh and India, which are idiosyncratic in their formations, physiographic characteristics, peculiar flora and faunal assemblages and in the behaviour pattern of resident animals. A huge population of over three million people depend directly or indirectly cn these forests.

India has only 6,81,976 hectares of mangroves forests of which 4,18,828 hectares occur in the Sundarbans and another 1,15,000 hectares is in Andaman and Nicobar islands. The remaining 1,47,888 hectares occur in the deltas of Mahanadi, Godavari, Krishna and in Saurastra and Kutch. Except in the Sundarbans, the forests elsewhere are in a deplorable state. This calls for immediate rehabilitation of these forests. The mangroves in Andamans have also suffered from heavy felling for collection of fuelwood.

The Dynamic Nature of Mangrove

Estuaries are complex biological sites which are very rich. Here the opposite currents meet and oscillatory tidal currents are formed which made the sedimentary system complicated. Physical, chemical, mechanical, and animate environments play a crucial role in modifying activities of microbial organism and other animate objects which are linked by processes resulting in a continuous exchange and assimilation of energy. The 'internal' process of energy fixation, accumulation of biomass, decomposition of dead material and mineral cycling are strongly

PHYSIOGRAPHY OF SEA COASTS OF BAY ISLANDS

Littoral Forests

(***Conspicuous species***)

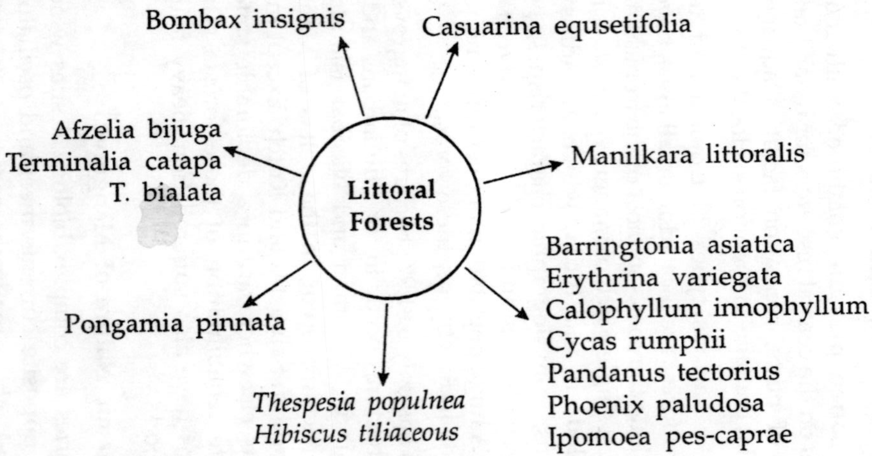

Chart 4

Conspicuous herbs of sea coast

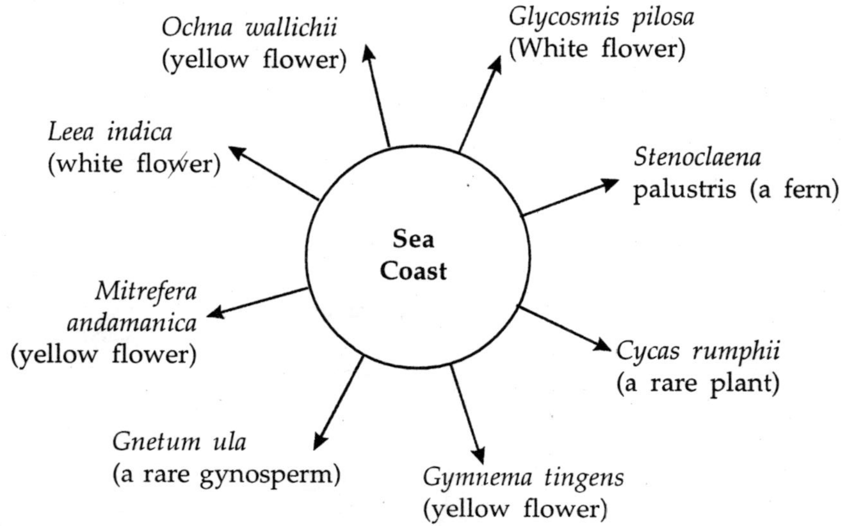

Chart 5

Conspicuous trees of sea coast

Mighty trees
Manilkara littoralis
Elaeocarpus robustus
Casuarina equiset ifolia

Trees of medium height
Pongamia pinnata
Ficus gibosa
Streblus asper

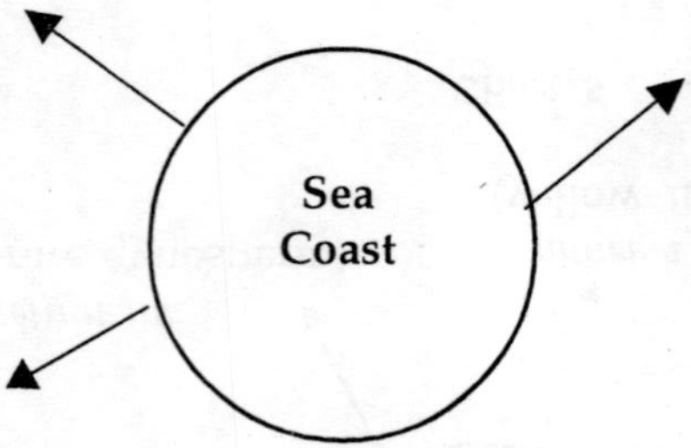

Mighty trees of adjoining island
Salmalia insignis (Didu)
Canarium euphyllum (white dhup)
Dipterocarpus Sp. (Garjan)
Pterocarpus dalbergioides (Padauk)
Pterocymbium tinctorius (Papita)
Terminalia procera (Badam)

Chart 6

Conspicuous plants of sea coast

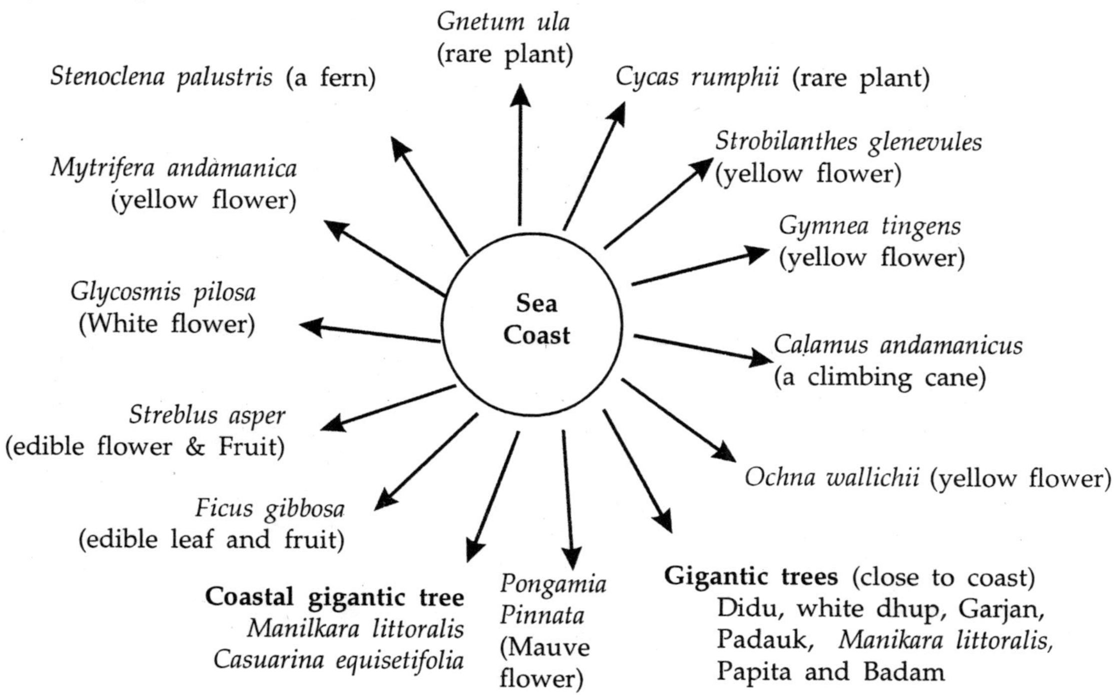

Chart 7

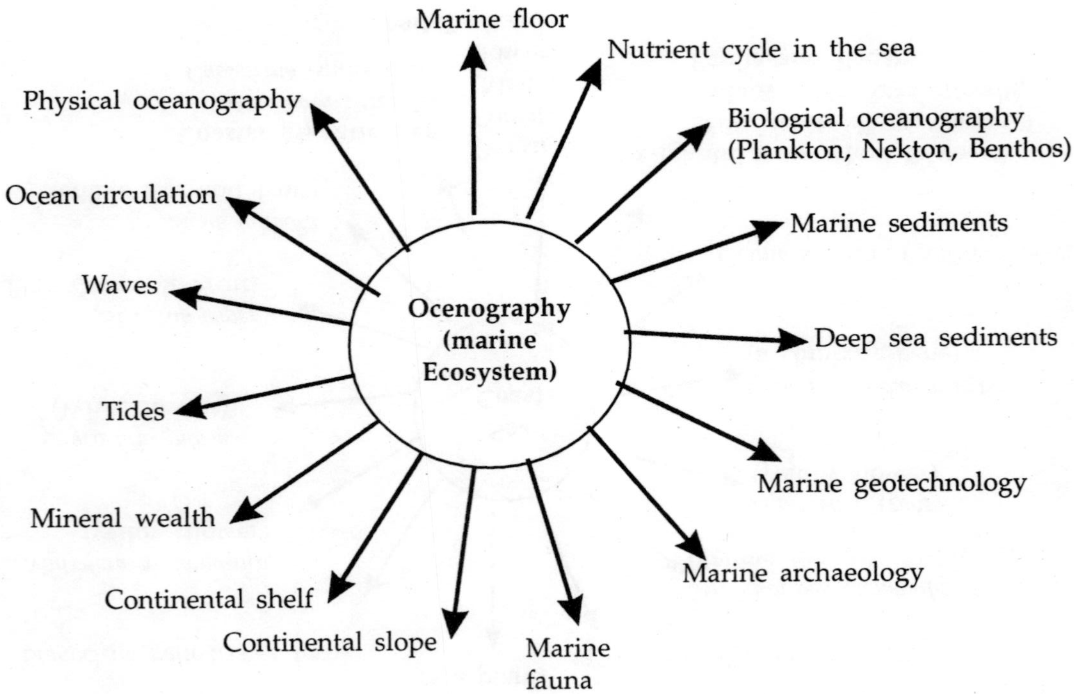
Ocenography
(marine
Ecosystem)
Marine floor
Nutrient cycle in the sea
Biological oceanography
(Plankton, Nekton, Benthos)
Marine sediments
Deep sea sediments
Marine geotechnology
Marine archaeology
Marine
fauna
Continental slope
Continental shelf
Mineral wealth
Tides
Waves
Ocean circulation
Physical oceanography

Chart 8

Diverse Sea Coast

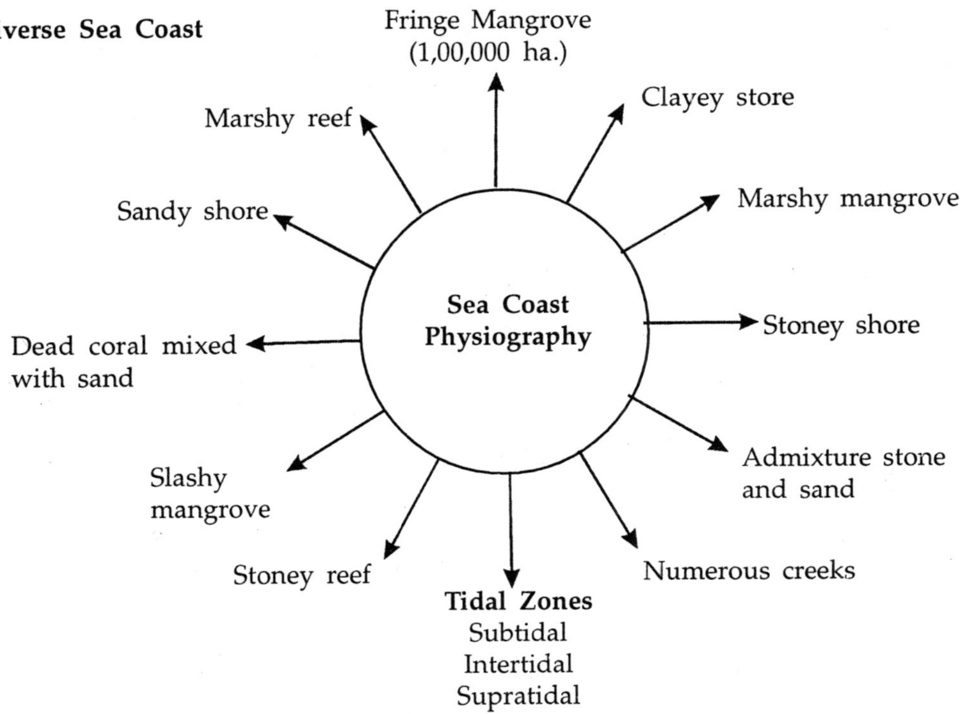

Chart - 9

PHYSIOGRAPHY OF SEA COASTS

Sandy coast

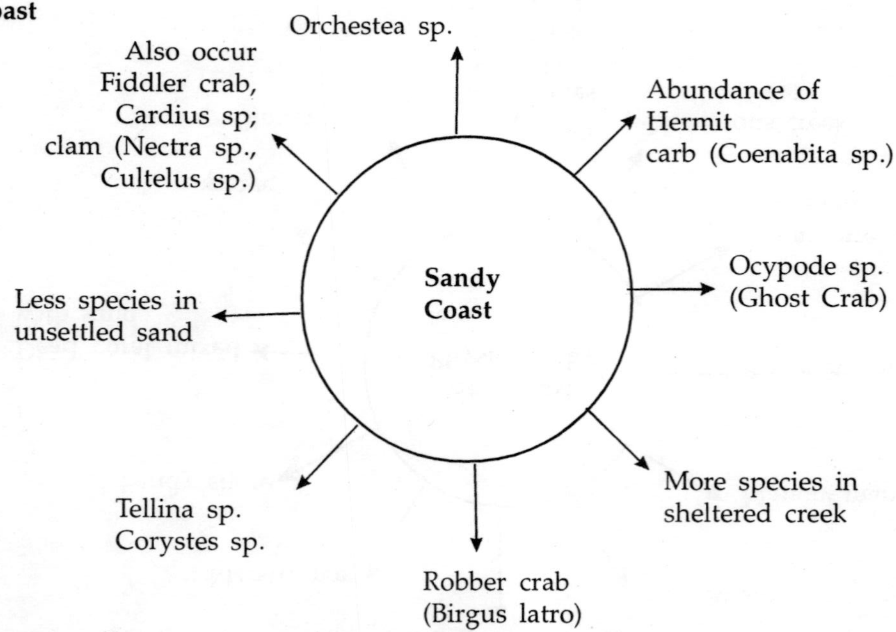

[Source : Das & Deb Roy, ZSI — 1989 for Species list]

Chart 10

Stoney Coast

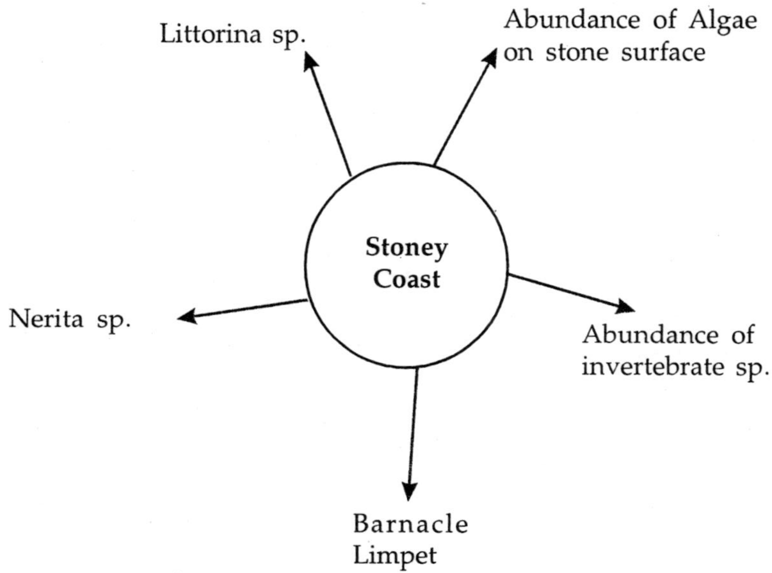

[Source : Das & Deb Roy, ZSI — 1989 for Species list]

Chart 11

Clayey Coast

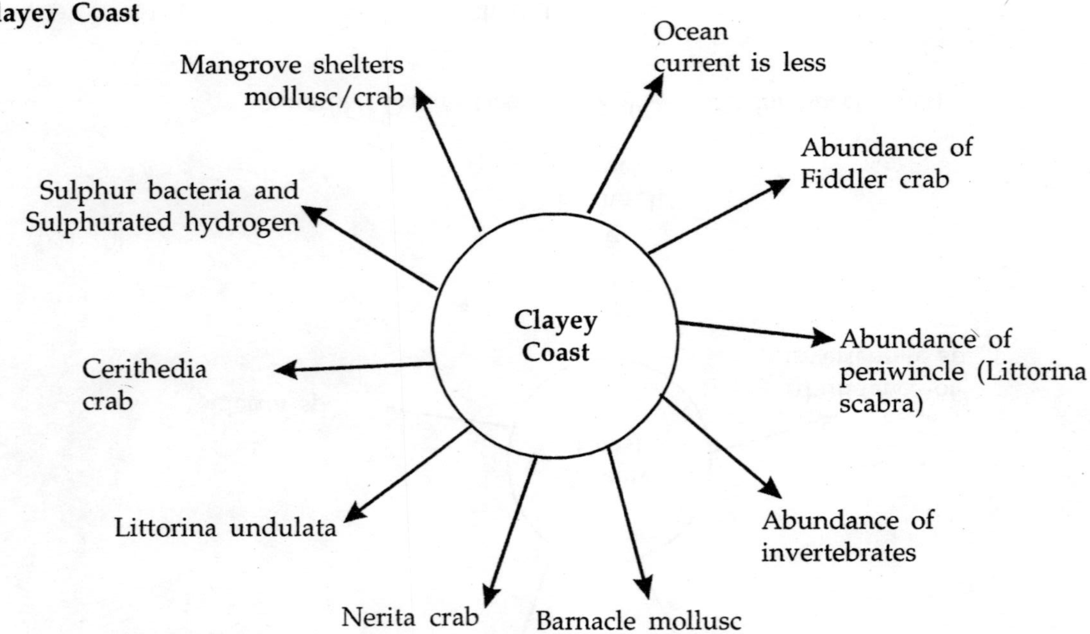

[Source : Das & Deb Roy, ZSI — 1989 for Species list]

Chart 12

(a) Some Bay Islands treasures are :
A

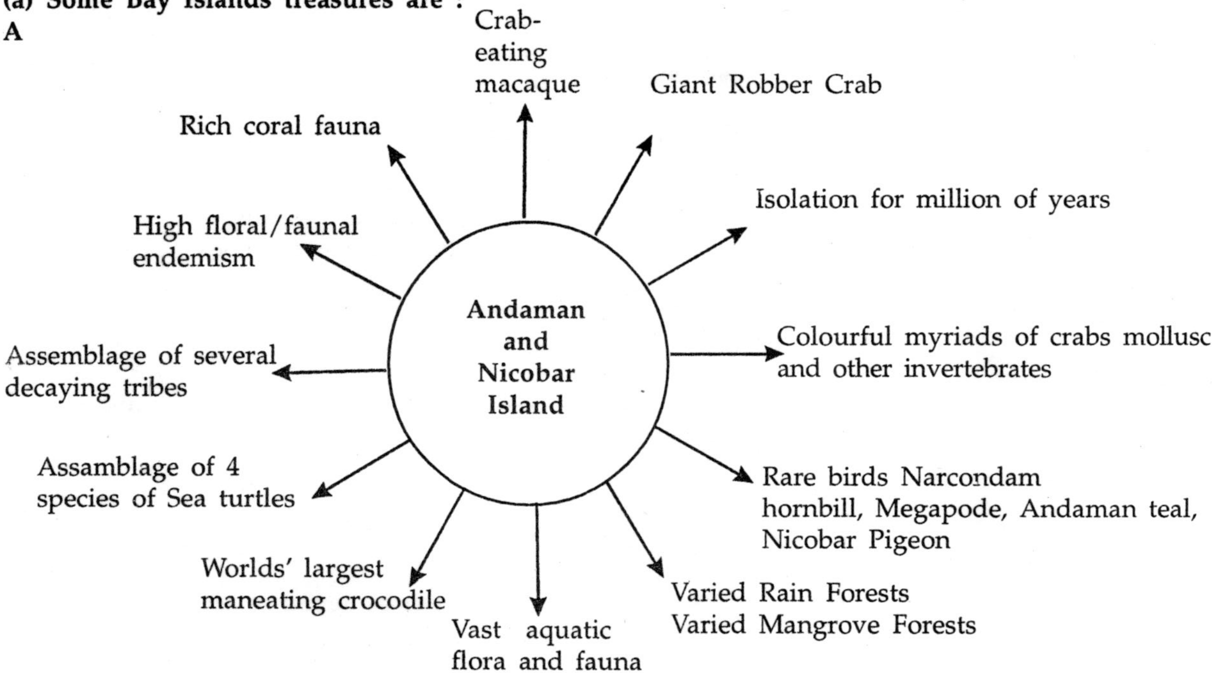

Chart - 13

B

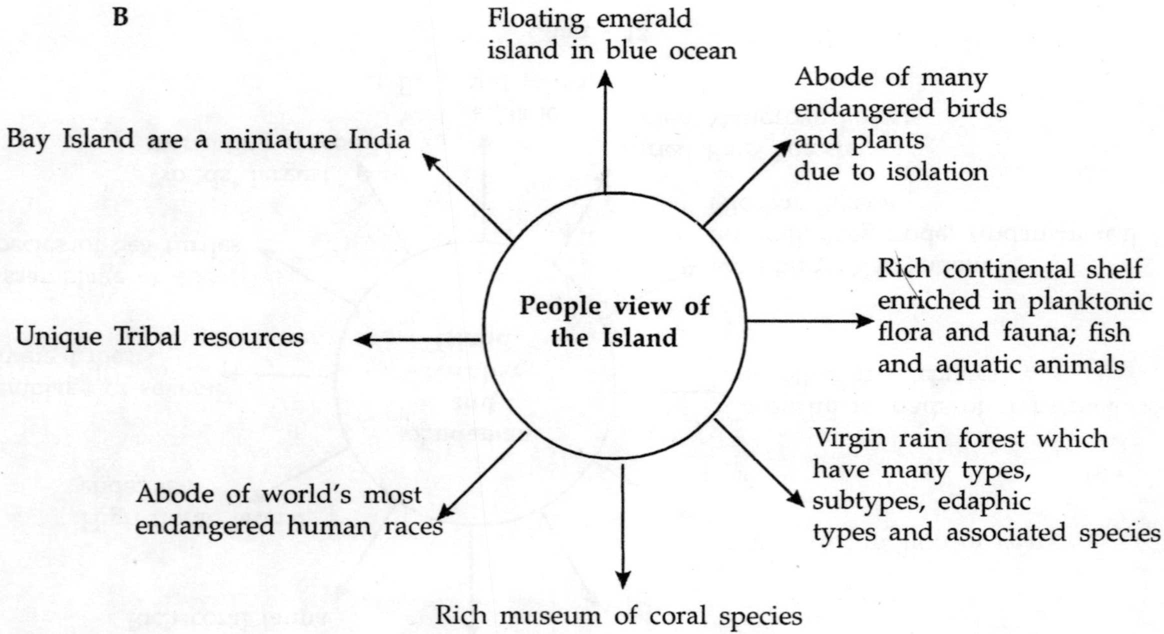
People view of the Island
Floating emerald island in blue ocean
Abode of many endangered birds and plants due to isolation
Rich continental shelf enriched in planktonic flora and fauna; fish and aquatic animals
Virgin rain forest which have many types, subtypes, edaphic types and associated species
Rich museum of coral species
Abode of world's most endangered human races
Unique Tribal resources
Bay Island are a miniature India

Chart 14

C

ANIMALS IN SEA COAST

Importance genera of Crustacea

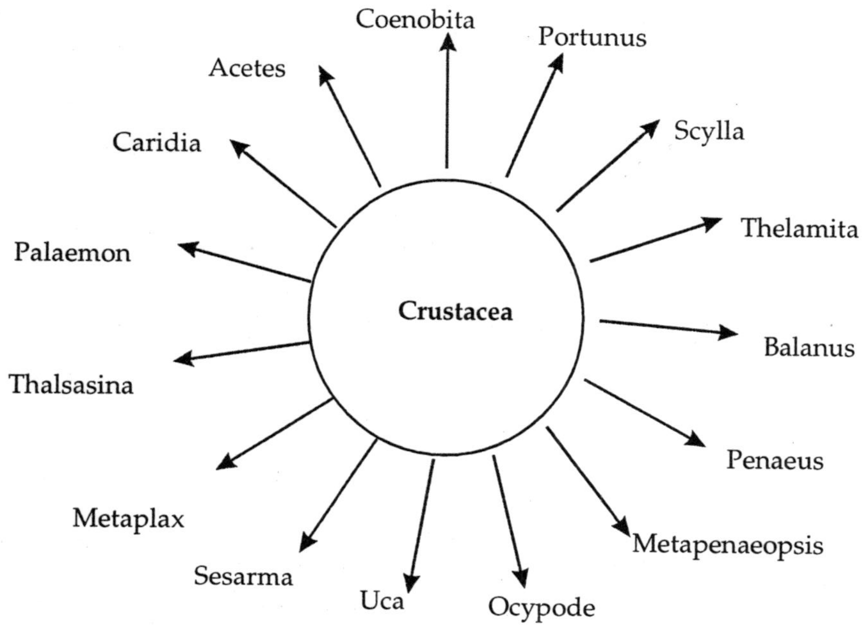

Chart 15

Hermit Crabs

- Intertidal sp.—Clibanarius longitars collected from the shell of Terebralia, Purpura, Achinata
- Supra-littoral sp.—Caenobita cavipes, collected from shell of Nerita, Cerithidea, Cerithium, Polynices, Clypeomorus.

These are a few genera and species.

influenced by a small set of critical 'eternal' processes which affect the availability of water and essential nutrients. The principal characteristics of the coastal environment are both cyclical and unidirectional because of intense interaction between land and sea. The meeting zones of fresh and sea water generate a ecosystem where nutrients are tapped. Food web structure in coastal zone is extremely complex. Benthic micro-algae and organic detritus are seat of algal micro-organism) constitute the principal energy source for vast population.

Mangrove Resources

Mangrove forests are characterized by a very low floristic diversity. There are about 90 plant species of which 55 are restricted to mangrove swamps. The indo-pacific region is very rich in number of species and the quality of these forests is very high. Excellent quality charcoal is produced from Rhizophora (Gurjan) which has a heavy demand. Firewood is produced from many species and is heavily exploited. Heritiera (Sundri) of Bangladesh is the most valuable timber of the mangroves. Excoecaria (Gnewa) is exploited in Bangladesh for pulpwood and matchwood; elsewhere it is used as packing case timber and as fuelwood. Xylocarpus (Pussor) is used for house post. Tannin obtained from Cerops (Goran) and Rhizophora (Gurjan) are used in leather, ink, plastic and formaldehyde industries. Various parts of mangrove plants are also eaten. The fruit of Avicennia (Bean) and Sonneretia (Keora) are edible as are the seedlings of Avicennia Radicles of Bruguiera (Kankra) are eaten. Salads are made from the leaves of Sonneretia (Keora) and from the herb Sesuvium (Gimasag) considerably quantity of honey and wax is collected from Sundarbans. Nipa (Golpata) is a very useful plant of the mangroves. Lower surface of young leaves is used as cigarette wrapper, while the older leaves are used for thatching purposes. The leaves also serve as raw material for paper industry. The sap of the inflorescence can be tapped for sugar, alcohol, vinegar production while its tapping gives a juice and a whisky is produced from it.

The mangroves support and sustain marine fisheries by serving as nurseries and as feeding ground for many marine invertebrates and fish by producing large amount of detritus for the marine food web. These forests serve as a link between terrestrial and marine ecosystem which is highly productive. Primary production is high as dead organic matter and detritus being consumed within the mangrove. The dead leaves are fed upon by carbs and other animals. As the leaves have high cellulose content they are only partly digested by the animals and are broken into small pieces. These broken pieces provide substratum for bacteria and fungi. This process of partial digestion and microbial growth is repeated which constitutes the basic for the detritus food web. Innumerable species of crustaceans, mollusks and fish take part in the degradation of dead plant tissues.

Part of the life cycles of may marine organisms is spent on the mangroves. Many species of Penaeid shrimps spawn off-shore using mangrove as a refuge and feeding ground in later stages of their life cycle. It is not known if the "mangrove dependent" species only require a bottom mud with certain salinity regime or are influenced by other factors. Shrimp and fish population migrate from place to place and it has been found that there is a positive correlation between shrimp landings and mangrove forests; also such correlation exists between landings of penaeid shrimps and areas of inter tidal vegetation. The large fresh water prawn Macrobrachium species which reach a length of 30 cm passes through their initial stages of the life cycle in brackish water and later wander upstream to fresh water areas. Mud crab-Scylla serrata, the delicious and highly priced crab, form a highly delectable item of food. Cultivation of this and other crabs in the Philippines, Thailand and in Singapore is done extensively by private agencies, mangrove oysters are highly marketable edibles. Various snails, clams, and cockles are collected from mud flats. Even the mud-skippers fetch high price in Hongkong and in the Philippines. In the Sundarbans there are more than 87 species of fish a good number of which form commercial species. The gulf of Mexico provides classic of estuary dependent fisheries. About 90 per cent of the commercial catch and 70 per cent of the recreational catch in the area are dependent on the estuarine habitat for part or the entire life cycle. More than 40,00,000 hectares of tidal areas in southeast Asia has been brought under brackish water pond culture.

Myrids of species and individuals of invertebrates, their specialized life-cycle, colouration, and behaviour pattern in

Special features of the Two Sites Sundarbans and Andaman Mangrove habitats

Sundarbans	Vast Marine Ecosystem in between the two sites.	Andaman & Nicobar Islands
- A Heritage Site - A Biosphere Reserve - Tropical vast Mangrove area.	Vast planktonic flora and fauna	- Isolated Bay Island - Tropical Rain forest - Mangrove forest
- Home of man eating tiger and crocodile.		- More than 100 wild-life Sanctuaries and National parks.
- Several Wildlife Sanctuaries and a National Park.	Vast number of Marine fishes.	- Biosphere Reserve
- A Project Tiger Reserve		- Tribal Reserve for a few decaying Tribes.
- Several endangered turtles and reptiles.		
- Rich Crustacea, Mollusc and planktonic floral and faunal zones	Sundarbans and Andaman Mangrove	- Many endemic and several endangered species.
- Very high tide level.		- Extensive Coral eco-system.
- Various types of coasts Sandy, Clayey etc.		- Man eating Crocodile.
- Protective & economic role	A vast area of Continental Shelf.	- Various types of coasts (Sandy, Clayey stones) - Protective & economic role

Chart 16

SPECIAL FEATURE

Salinity

Various gradation -

- Exudation excess Salt by some plants.
- Some has salt glands.

Mangrove Habitat

- Gall like formation helps exudation of excess salts.
- Various types of roots - Knee, Buttress, Pneumotophores, Stilt.
- Crassostrea Cuculata's attachment to mangrove tree trunk.
- Preponderence of Hermit Crabs.
- Osmoregulation mechanism.

- High salinity effect Plant growth
- High salinity may cause towering the transpiration.
- Heritiera forms better in less Salinity in soil.
- Some species characterized by Viviparous generation.
- Air breathing Terrestrial (Amphibian) fishes.
- Common molluscan sp. are Telescopium and Cerithidtea sp.

Chart 17

coastal and sandy shore are specific qualities of the mangroves. Of the mammals the occurrence of man-eating Royal Bengal tiger and their behavioural pattern is unique anywhere in this world. Pigs and spotted deer also occur in abundance. Estuarine crocodiles (Crocodylus porosus) are the largest and mot ferocious of all crocodiles. Monitor lizards, rhesus monkeys, fishing cats, dolphins and sharks in water are few, amongst many, animals of the mangroves. Proboscis monkeys are endemic to Borneo. Crab eating macaques, Presbytes langur and flying foxes are other animals of the mangroves. In Papua New Guinea spotted Cuscus are very common. Of the birds, few are endemic to the mangroves. In the sea coast and in inner marginal land large number of local migratory birds aggregate in breeding season which include cormorants, darters, herons, egrets, whimbrels and many other marsh and water birds.

Economic Consideration of Land Use

Most of the mangrove dwellers earn their livelihood from fishing, honey collection, fuelwood and timber collection and shell collection. In Thailand over an area of 200 sq. km. where 480 households are engaged in catching fish, prawn, shrimp, crab, squid and shell annual catch was valued at 1.3 million in American dollars. Shrimp and oyster cultivation in Thailand yield about 17 tonnes per hectare per year valued at 31,500 dollars. Mussel yielded in this country is 180 tonnes per year per hectare. Earning from cultivated sea weed in China and Korea is high. In the Philippines a sea weed farm of 0.5 hectare had provided a net income of 1360 American dollars. In some of the south-east Asian countries milk fish and shrimp catch from intensively managed area were over 200 kg to 3000 kg per year per hectare.

Mangrove forestry, capture fisheries and open water aquaculture are largely complementary activities which are one form of land use. Removal of mangrove forests adversely affect capture and open water aquaculture. This indirectly means disappearance of mangrove dependent species. The highly productive and dynamic sites can be very usefully utilized by judicious management to uplift the standard of living of the poor mangrove dwellers. The following will testify to high potential of mangrove sites.

Charcoal, Nipa thatch, small scale fishing, oyster cultivation and shrimp cultivation may yield value of 318, 230, 39, 60 and 206 American dollars per year per hectare respectively; in a commercial farm shrimp cultivation has yielded a sum of 2106 American dollars per hectare per year where rice cultivation in the mangroves yielded an income of 165 American dollars.

In terms of gross annual income per hectare the best land use appears to be intensive shrimp cultivation. Charcoal also yields a very high value (a plantation yielded a value of 400 American dollars per hectare per year).

All these indicate that there must be judicious management of the productive mangrove site as all the processes and animate objects are intimately connected with each other.

Conservation and Human Welfare

Besides utilizing the resources for monetary gain there is urgent need to protect the forests from conversion to agricultural land and fish-prawn breeding sites. Sincere effort, however has

Speciality of the Study Sites

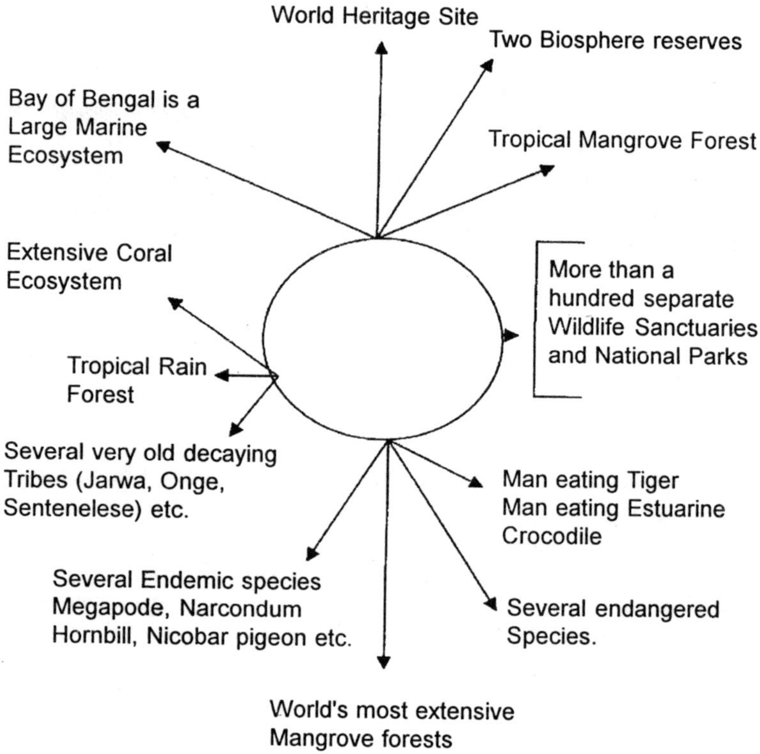

Chart 18: Sundarbans and Andaman Nicobar Islands

been made to preserve the mangrove ecosystem as a whole in Australia where 23 'Habitat Reserves' have been marked covering an area of 93,925 hectares; but mention must be made to Kakadu National Park in northern territory which gives protection of about 614,400 hectares. Bangladesh and India afford protection of about 150,000 hectares of mangrove forests in the Sundarbans. Brazil, Colombia, Costa Rica, Dominican Republic, Malaysia, Mozambique, New Guinea and many other countries have felt the necessity of preserving such forests. These sites have provided immense recreational and aesthetic facilities to the people, besides affording research facilities on unknown plants and animals and on their behaviour pattern. The utility of the mangrove forests to counteract the force of cyclonic wind and waves has been felt in India, Bangladesh, in the Philippines and else where. The destruction of these forests may eliminate very many species of fish, shrim and prawn besides other species which spend either entire or part of their life cycle in the estuaries. Acid sulphate that is produced in soil where tree felling has been done is indicative of misuse of the mangrove productive sites.

Yet extensive destruction has been done to such forests, 1300 to 1500 sq. km. of forests have been reclaimed in Indian Sundarbans in the last 100 years. More than 40,00,000 hectares of tidal area in Asia has been brought under brackish water and pond culture. About 27000 hectares of mangrove land in peninsular Malaysia has recently been identified for brackish water fish culture. Biotic interference has depleted the mangrove forests all over India except the Sundarbans.

The foresters, technologists and economists must be involved in preparing master plans for best utilization of the mangrove forests. It is indeed a very difficult task as to how much area is to be set apart mangrove species by others more tolerant of the increased salinity. This can have severe consequences when a local industry is dependent on a sustained supply of the species being replaced, for example, the commercial mangrove forest tree, Heritiera fomes in the Gangetic delta of Bangladesh.

Reduced dry-season fresh fauna dependent on both a source of fresh water and sufficient food during the dry season. The aggravated effects on the Bengal tiger (Panthera tigris tigris), for example, have been documented, and similar information is available for other large animal residents of Asian mangroves. In the past in many tropical countries, natural mangrove forests have provided sanctuary for many species, simply due to their existence in a coastal landscape otherwise modified by economic development.

High water salinities affect mangrove-dependent fisheries. The reduced flushing may result in the accumulation of detritus where it is unavailable to dependent offshore communities. Too many barrages constructed on the Indus has resulted in less discharge of freshwater for about nine months as a result of which mangroves have vanished.

Bangladesh

The mangrove in the delta of Ganga-Brahmaputra has been in a state of progressive deterioration. Salinity gradient controls the structure and composition of forests as may be seen in the Bangladesh (less salinity) and West Bengal (high salinity). Inspite of the fact that the Sundarbans Forest is a major source of round wood and pulpwood in Bangladesh and is a major wildlife refuge for certain rare and endangered species (e.g., the Bengal tiger Panthera tigris tigris), only minor attention has been given to the deterioration on the forest. Further, it is doubtful that continuing economic development based on fresh water resource development will be sacrificed or altered to protect the Sundarbans. However, co-operation between India and Bangladesh has resulted in a greater dry-season flow of freshwater to south-western Bangladesh.

Forest Exploitation

Mangrove forests are much used for fire wood (India & Africa), Charcoal (Thailand, Malaysia and Indonesia). Timber is also exploited all over for construction, pulp and paper. Large scale settlement of people by felling mangroves has been done in Venezuela and East Kalimantan (Indonesia). Large scale deforestation has also been made in Sarawak, Sabah, Matang, Venezuala, Sulawesi, North Sumatra and other areas.

Conservation to Agriculture and Aquaculture

Coastal swamp forests in saline, anaerobic environments have traditionally been considered to be marginal or totally unsuitable for agricultural and/or aquacultural production. The Rice Research Institute in Los Banos, Philippines find ways to utilize this marginal resource which includes not only mangrove dominated lands but also contiguous areas of tidal flats, beach ridges, inter creek, saline flats and reclaimed lands. In Asia and Africa the pressure in arable land has led to conversion of mangrove forests into agricultural land. This has been done in Senegal, Gambia, Sierra Leone, India and Bangladesh.

Aquaculture projects have expanded enormously in recent years resulting in conversion of 1.2 million ha of mangrove forests (1977). In Philippines aquaculture area was 174, 101 ha in 1972.

As of 1981, there was demand for additional 70,300 ha of mangrove areas. Ecuador and Costa Rica mangroves have also suffered such consequences.

Coastal Development

Coastal development for tourism, urban area extension etc. has taken place in Australia, New Zealand, U.S.A. and in other countries. Singapore has also suffered a lot of reduction of area.

Salt Pond Construction

Salt ponds are built on mudflats or more commonly in mangrove areas. In Gujarat there is a huge Chemical-Salt Industrial complex. In Benin, Malaysia and some other countries there are salt pond areas.

Conservation of Sundarbans Mangrove

This is a vital issue. Experts have felt the utility of mangroves in combating a disaster like "Tsunami".

Sundarbans mangrove conservation is vital to save Calcutta and millions of coastal people from the evil effect of global warming and hazards like 'El-Nino' besides the fury of cyclone, flood and the like.

PART I

TROPICAL FORESTS OF SUNDARBANS MANGROVES

(FLORA, FAUNA, MAN-ANIMAL INTERACTION AND CONSERVATION)

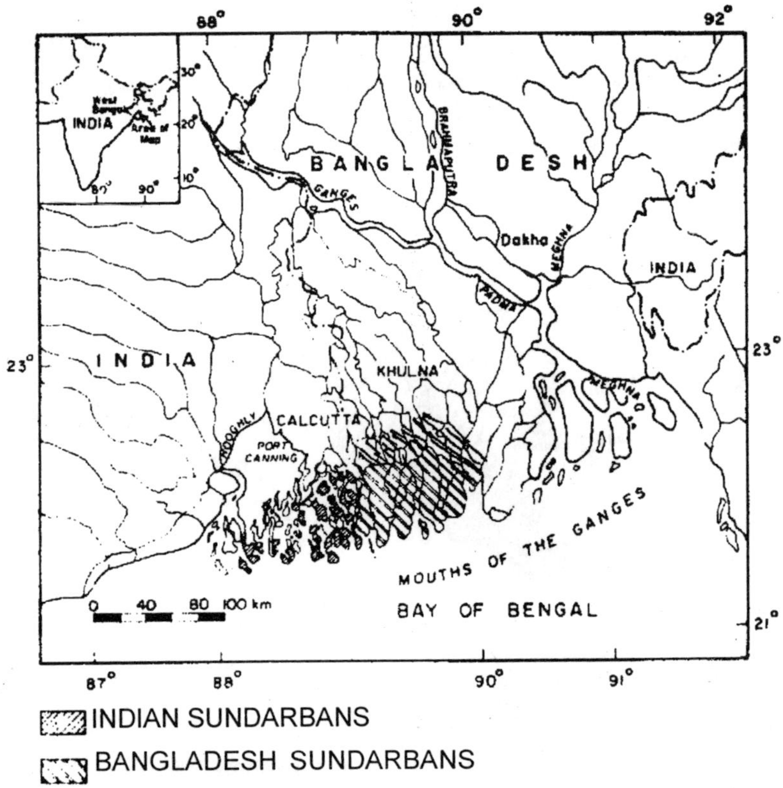

MAP 2. The Sundarbans (after Blasco, 1975)

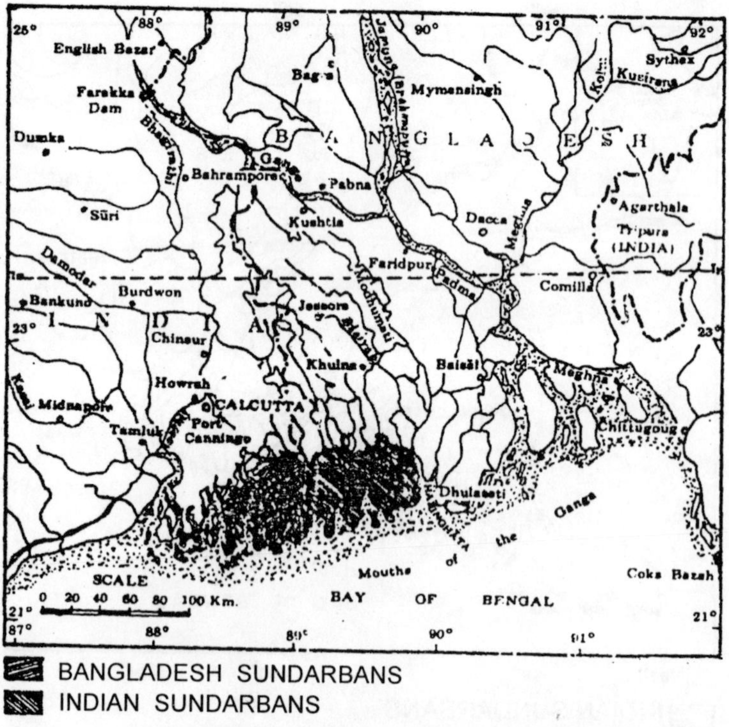

Map 3: Delta of Ganga and Brahmaputra

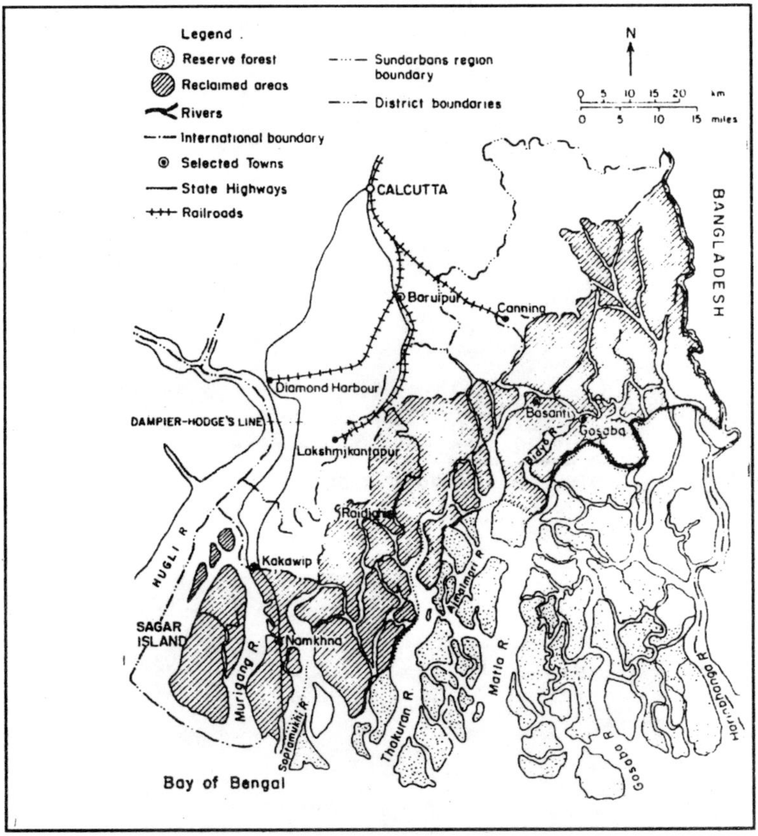

MAP 4. The Indian Sundarbans

SOME FLORA, FAUNA AND FEATURES OF SUNDARBANS IN PHOTOGRAPHS

(Courtesy - Forest Deptt. West Bengal)

Fiddler crab

Tiger prawn seed collection using mosquito net

Tidal creek

Black capped Kingfisher

Lord of the Sundarbans

Amphibious Tiger of Sundarbans

Crocodile breeding at Bhagabatpore

Rearing Olive Ridley Turtle at Sajnekhali

Pnumatophore - a unique feature of Sundarban

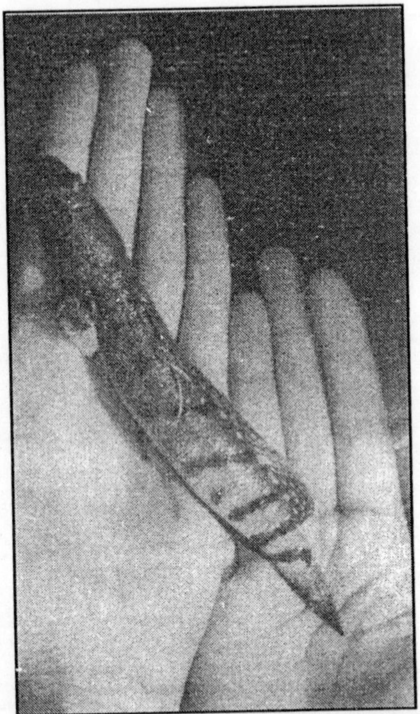

Mud skipper

Water monitor lizard

1

Flora and Fauna in the Abode of Man-eater's in Tropical Mangrove Forests of Sundarbans

THE TERRAIN

Situation

The Sundarbans covering some 10,00,000 ha of land and water are part of one of the world's greatest deltas formed by sediments deposited by three great rivers the Ganga, Brahmaputra and Meghna. Almost 62 per cent of the Sundarbans is situated in Bangladesh, while the remaining 38 per cent of the regions, lies within India. These two portions together comprise the largest continuous block of mangroves in the world.

Indian Sundarbans are located between 21° 32′ – 22° 40′ north and between 88°85′ – 89° east. The region is bounded by the river Hooghly to the west, the Bay of Bengal to the South, the Harinbhanga – Raimangal rivers along international boundary with Bangladesh to the east and the Dampier–Hodges line to the north. There are 2151 rural settlements, 102 islands; 54 islands being inhabited. The population is 5.8 million (1991 census). Urban centers/Towns comprise 22 nos.

The Sundarbans play an important role in the lives of the people in the neighbouring communities who depend on the forest for subsistence and protection from natural calamities. The forest also supply raw materials for a number of industries. Direct and indirect income generation through the harvest of product, primary and secondary processing, transportation and marketing is substantial.

The forest was administered as a single management unit (Forest Management Division) until 1947, when India was partitioned. The Indian Sundarbans are now administered by the Forest Department, Government of West Bengal.

2

Socio-cultural Scenario (Human Settlement, Agriculture, Culture and Religion and Tiger Cult)

The visitors will appreciate the problems of poverty stricken people who suffer from absence of some basic requirements of a civilized society. For the tourists a few most important aspects have been discussed to give them an impression of the complex socio-economic status of this complicated terrain.

SETTLEMENT PATTERN

Situation: 21° – 32′ to 22° – 40′ North Latitude

88° – 05′ to 89° – 00′ East Longitude

Area: 6,388 sq. km.

Population: 5.8 million (1991 census)

Rural settlements/villages: 2151

Urban centers/Towns: 22 nos.

Of the 22 blocks, 8 comprise the famous tidal Mangrove forests, which are infested by wildlife, bounded by Dampier – Hodges line.

Survey of Sundarbans: The area was meticulously surveyed by the Britishers braving the hazardous inaccessible, inhospitable, dangerous and terrible terrain. They successfully did the mapping of the same. A brief chronological description is furnished below.

- Hooghly river to passar river survey was carried out by Lt.
- W.E. Morrison during the period 1811 to 1814.
- Hooghly river to Jhakura, surveyed by Lt. Blane (1813-14)
- Princep surveyed Blocks (lots) in 1922-23.
- Lt. Hodge surveyed the boundary of forests in 1929-(Morrison map) – Hodges map of Sundarbans, 1931.

In 1929-30 Dampier, the then commissioner and Ltd. Hodge, Surveyor surveyed and defined the limits of forests – Dampier-Hodge Line.

SETTLEMENT PATTERN

The Sundarbans delta provides a micro level diversified picture amidst the overall monotony. There exist various types, shapes and forms of settlements.

The over-shadowing influence of the Calcutta district is a major hindrance, lying in the way of the development of the settlers.

Among Inter-regional pattern of human settlements, the following are the major facets of the settlements of this region.

- Economic Profile of Household.
- Social Profile of Household.
- Mortality, Morbidity and Health Care.
- Family Planning and Health Care.
- Quality of Housing of the Human settlements.
- Basic amenities and personal hygiene.

World Summit on Sustainable Human Settlements was hosted at Istambul in June, 1996, while many areas of Sundarbans still dwelling in the realm of backwardness and poor quality of settlements and housing.

In this context for sustainable development of Sundarbans areas following Action Plan is suggested: -

- Strengthening of the earthen embankments by masonry works.
- Fortifying of vulnerable embankments by exclusive masonry structures.
- Installation of hi-tech cyclone warning system, well equipped with trained staff along with elaborate modern communication facilities and infrastructure.
- Creation of adequate numbers of shelters for victims of cyclone/storm.
- Making provision for better housing facilities. Elaborate and extensive network of water supply arrangements for supply of safe drinking water.
- Extensive electrification of the villages, charging subsidized rates per unit consumption.
- Improvement of communication system by introduction of adequate no. of power boats and hovercrafts for ferrying people from place to place.
- Radical improvement of education network for imparting compulsory education to the people from all walks of life.
- Extension of medical facilities.
- Extension of eco-friendly tourism at different salient spots.
- Prevention of illicit collection of fuelwood and poaching.

SOCIO-CULTURE AND CULTURAL PROFILE (A BRIEF OUTLINE)

Folk religion revolves in Worshiping of village God.	Folk rituals and rites
Fertility Cult Phallic Cult	Terrain (broken and interspersed by rivers, creeks and kharis, affected by high and low tides, Cyclones, storms and inclement weather).

Totem worship Magic	Cropping in adverse conditions Agriculture and horticulture.
Demographic issue	Special type of forest, trees and animals (inland and water).

Principal Folklore deities are:

- Dakshin Rai
- Kalu Rai
- Badar Sahab
- Bakra Gazi
- Sonapir
- Sawal pir
- Gazi Sahed
- First Goddess of forest is Banbibi
- Maa Manasha is worshipped to save people from snakes

Cult of ancestors

Woodcutters, Honey collectors and Fishermen worship:

- Dakshine Rai
- Narayani Maa
- Banabibi
- Gazi Saheb
- Kalu Rai
- Barkhan Gazi

SOCIO ECONOMIC STATUS

Economic status of the people of Sundarbans

- 85% of the population depend on agriculture in reclaimed lands.
- 50% of the agricultural labourers are landless.
- 44% of the total population belong to S.C. and S.T.

MAN-ANIMAL CONFLICT

- Human population pressure exerts stress on the habitat, resulting in its deterioration.
- People, being poor, have to depend upon the forest and its resources.

RESEARCH ON VARIOUS FACETS

Several anthropologists and socio-economists have studied the problems of this terrain and about how the standard of living of the people of this region can be improved. In recent years, groups of experts have worked in various socio-economic facets of this region, which are depicted in the following:

- M.K. Sandhu et al., on growing Vegetables followed by Aman paddy.
- A.N. Bose on Brakish water and Marine Fisheries.
- A. Sengupta on Economic Activities, industrial and Economic growth.
- Sharmila Das on fishing environment and fishing practice etc.
- N. Chatterjee on Impact of Irrigation on landuse, cropping Pattern & Crop yield.
- G.N. Saha et al., on Settlement of the reclaimed portions of Sundarbans.
- Chatterjee and Bagchi on Transport System.
- Anuradha Banerjee on Relationship Between the natural and Human resources and the development Process.
- N.R. Kar on Economic character of the surrounding of Calcutta.
- D. Basu on Pattern of Settlements in Sundarbans.

ENVIRONMENTAL HAZARDS ON HUMAN SETTLEMENTS

During pre and post monsoon seasons the lands in the coastal belt of Sundarbans witness 2-3 severe cyclonic storms every year. The fury of the associated wind-force, high storm surge and tidal bores cause large scale devastation and disaster by way of loss of human lives, damage of dwelling houses, shelters, crops and forests as well as damage and destruction of embankment, which result in acute scarcity and shortage of drinking water, extensive damage to agricultural crops and also the land. Extensive damage is caused to the fishermen and their boats. 91% of the families get severely affected by storms and cyclones.

TIGER CULT

Banbibi

The lady of the forest is the presiding female deity of the Sundarbans cultural zone. She is the guardian deity of the forests. Both the Hindu and Muslim communities pay their respects to her before venturing into the forest. Also known as Basuli, Bisalakshmi, Bon Durga or Bon Kali. She is portrayed by clay modelers either mounted on a tiger or a hen. She is pretty and graceful, ever eager to protect the people of the Sundarbans. The deity is often dressed like a Muslim woman with plaited hair and wearing a ghagra-choli with a brocade cap on her head. She carries a child in her arms. In some areas, the deity wears a sari.

Dakshin Rai cult

The importance of Dakshin Rai is significant from a social, historical, cultural and anthropological perspective. In different parts of south 24-Parganas he is depicted in two different iconographical forms, both made of earth, one depicting him as a clay model of a human in the likeness of warrior, dressed in yellow robes and armed with sword, gun, bow and arrows; and the other depicting him as a terracotta human head with large eyes, a moustache and leaf-like, conical head-gear.

Dakshin Rai is the god of the tiger. He is worshipped below a banyan (Ficus bengalensis) or neem tree (Azadirachta indica). Wood cutters, boatmen, honey collectors and other communities worship this god especially on Makar Sankranti day. Dakshin Rai legends sometimes depict him as a warrior seated on a tiger with a bow and arrow in his hands. He undoubtedly the main folk deity and his worship is an important ritual of the cultural and religious history of the Sundarbans.

Among the folk rites associated with the worship of Dakshin Rai, Jatal has a special place. This ceremony is observed in the forest at the dead of night. The worship is performed by lighting torches, beating of drums and offering of wine, meat and hallucinatory drugs. Ducks and goats are sacrificed. In these folk societies various beliefs exist about the divinity and prowess of Dakshin Rai. Invocation of Dakshin Rai is also made in case of diseases of cattle. Dakshin Rai is worshipped mostly as the god who linerates humans from evil influences and helps cultivation.

Iconographical studies of the chopped head symbol of Dakshin Rai have established that it signifies the primitive custom of human sacrifies and head hunting. The stories of Barkhan Gazi, Bonbibni and Dakshin Rai probably spread during Muslim domination in southern Bengal. In folk stories there are references to the strife among the religious lords and the clashes between the Hindus and the Muslims and of eventual communal harmony and cultural synthesis.

3

Man-eating Tigers - A Socio-cultural Issue

A visitors first requirement is to know about "Tiger God" or Maneaters of Sundarbans. Knowing fully well all about this supernatural animal, they should proceed to study this golden land. Some basic information are presented to the tourists to digest the most difficult problems of this land.

MAN-EATERS – VARIOUS CONCEPTIONS

The Sundarbans tiger stands supreme in the tiger map of the World. It is believed that they are hereditary man-eaters, fiercest, most cunning and intelligent of all tiger. To present observations include their habitat types, vulnerable operational zones, comparative pattern among different professional men, fatality in various age groups of men, monthly and yearly pattern of fatality, breeding potential, food habits and other aspects of their behaviour. Such studies have never been made in the estuarine tract in the present manner and it is expected that this study will enable to throw light, on many facts so far unknown about the tigers of the tract. The readers should refer to bibliography for further reference.

Nature's masterpiece of creation are the Tiger of Sundarbans. The resourcefulness, intelligence, agility as personified in such tigers have turned this magical species into a legendary myth. Superstition, and supernatural qualities attributed to the man-eaters, however fictitious they may appear, have their basis in fact. Locating its human prey, silently stalking him in ever closing circles for hours, high degree of intelligence to baffle trap gunner and trappers, diabolical understanding of the human mind, high precision in attacking human beings and crossing rivers, great power in controlling reflexes in mounting a boat, with least rocking, have made it a mythical animal. Blessings are invoked form Dakshin Rai (tiger God), Narayani Ma, Banbibi, Kalu Khan, Sa Jungli, Gazi Sahib prior to entering the forests as it is thought that the animal has supernatural qualities of destruction of human beings who are helpless against its machinations.

Tigers in the Sundarbans taken a toll of about six hundred human lives in sixteen years during mid decades of 1960s. These are all reported cases; it is difficult to estimate the nukmber of unreported cases. The victims are almost the most backward in the country. Sheer economic necessity compels them to brave the hazards to the terrain over and over again. Fishermen, honey-collectors, wood-cutters and others engaged in work of an allied nature constitute the bulk of "permit-holders". These are the people for whom the Sundarbans tiger is a living entity-almost a part of their lives.

When such people talk of their experience - no matter how tragic these are – the references to the tiger are rarely, if at all, in inimical terms. The tiger is "accepted". It is almost as if there is an elaborate ritual of ground rules to ensure co-existence. A priest accompanies them to the forest and invokes the blessings of Dakshin Rai – the principal folk deity and lord of the forest, in the robe of a warrior, armed with sword, gun, bow and arrows, and seated in a tiger - and also of Narayani Ma, Banbibi, Kalu Khan, Sa Jungli and Gazi Saheb to obtain immunity from any attack by Man-eaters, Crocodiles, Manasa (snake) and other creatures.

The naturalists will find this chapter immensely interesting and would be lured to visit Sundarbans and know about the rich flora, fauna, natural phenomena and the people of the area. They will appreciate how sincerely the foresters have protected the area over the years braving the hazard of man-eating tiger, man-eating crocodiles, man-eating sharks, venomous snakes, menace of international gangs of dacoits smugglers the natural hazards of cyclone.

AUTHOR'S REVELATION

Neither the forest nor civil administration bothered about human fatalities except merely noting down the number of people killed by tigers in the annual report of the division. The present author saw a freshly killed dead body of a permit holder in a fishing boat during his maiden tour of Sundarbans in 1970 as the DFO. Of the three persons in the boat, the grand father (64), the son (42) and the grandson (18), the son (42) was killed; his jugular vein was pierced and the death was instantaneous.

The author spent several sleepless nights thinking seriously to do something to reduce the fatality. The instructerd his field staff to interview hundred of permit holders and gather information about the behaviour pattern of the man-eaters in killing their prey - man and animal. This was done and several hundred true stories with countless encounters of the permit holders with the man-eaters in the forest were gathered. Of these stories the author selected 21 representative ones that revealed the behaviour pattern of the man-eaters and also the reaction of these permit holders from the mighty man-eaters.

ANALYSES OF VARIOUS PARAMETERS OF FIELD DATA

Analysis of the above data exhibit that although water salinity tends to increase as one proceeds from the Northern zone to the Central zone and then to the Southern zone of the Reserve, similar pattern was also noticed so far as soil salinity index is concerned, but the frequency of human casualties does not vary according to such zonations. This indicates that the man-eaters of the Sundarbans are concentrated in some habitat formations only, which also incidentally records significantly more salinity. It requires further probe to come to any definite conclusion to ascribe reasons for the man-eating urge of the Sundarbans tigers. It may, however, be concluded for the present that the man-eating urge, a peculiar trait marked in a percentage of the Sundarbans tiger, is governed by cumulative action of physical and chemical environments of the Sundarbans. It has further been noted that the abundance of scarcity of prey animals are not directly related to the human casualty figures to tiger.

The Sundarbans is a natural laboratory. It provides astonishing examples as to how a physical and chemical environment can exert influence to the different forms of life. The extent of endogenous and exogenous periodism, osmoregulatory adaptations amongst different organisms were perhaps nowhere so much pronounced and diversified as they are in the Sundarbans, man-eating tigers-astonishing and ferocious-world's largest and most ferocious crocodiles, water and land lizards, crab-eating spotted deer, monkeys and wild pigs, innumerable varieties of crabs and other wildlife, and more than two hundred species of fish etc., make the area unparallel.

Further detailed scientific study is likely to throw light into the hitherto unknown biological properties of the Sundarbans.

The Sundarbans tigers are branded as man-eaters since time immemorial, but the entry of government personnel deep inside the forests has not diminished. Men and tigers are living together through ages despite cases of human fatalities. The problem of man-eating in the Sundarbans is thus, not, something which bewilders the people, as they accept these incidents in their life-style and a living system. The problem of man-eating elsewhere can hardly be compared with the man-eating incidences in this tract. Jim Corbett used to attribute old age, disease and injury to man-eating in the tigers. This does not seem to be very cogent reason of man-eating in the Sundarbans tigers. The most swampy terrain of the Sundarbans with incredible plant density with sharp and pointed breathing roots. The physical and chemical environment of the Sundarbans obliterates all limits of land and water animals who are forced to lead amphibious mode of life. In this peculiar eco-system the intruding man becomes the easiest available prey to the adventurous tigers, as he is defenceless in the face of any attack from the tiger. The Sundarbans tigers with diabolical understanding of human nature keep a close vigil on the seemingly dwarf creatures of men while they move in sitting postures in boats and dinghies; Intake of saline water and the soil also is likely to create physiological imbalances not only to the tigers, but also to other wildlife denizens; there are horizontal and vertical stratifications in salinity indexes in the terrain of the Sundarbans and the indexes also maintain seasonal fluctuations. Thus it is not correct to conclude that all wildlife denizens of the Sundarbans should be affected to an equal extent by salinity, as same are likely to be affected more than the others, according to their limits of tolerance to variants and the vagaries of the eco-system. People entering the tigerland of the Sundarbans hardly take any precaution against these man-eating tigers because of their utter callousness, tradition and, above all dire poverty. People go out in open dinghies and boats for months together to eke out a miserable living for them and to support their dependents and fall an easy victim to the circumstancial wily man-eating tigers. This is deemed as a normal fact of life and is sadly accepted as occupational hazard. Payment of adequate compensation in time may also improver the man-tiger interactions to a very great extent. Concerted and all round fringe area development programme may remove the biggest human pollutant, the poverty, in a big way and successful implementation of the same is likely to pay rich dividends in improved man-tiger inter-actions in the terrain.

Some peculiar innovative attempts in electric fencing and battery charger were experimented in the terrain of the Sundarbans sporadically without monitored observations and the protagonists have claimed loud successes from these experiments conducted for the reduction of human fatality. The protagonists, for the time being forgot that man-eating tigers do not restrict their killing operations only to official permit-holders since a large number of persons entering the area without official passport also get killed by the tigers and the success stories do not include such cases who are branded as illegal entrants. Success or failure of an experiment pertaining to reduction of human casualty cannot be judged from the figures of official entrants only, which at best only tell half the story. So, if there is anything called priority in the delivery of human security system, these poor entrants to the Sundarbans need be provided with a safe sailing boat or dinghy rather than unmonitored and casual experimentation on electrical fencing or battery operated mechanisms. They may be provided with awareness and a sense of security plus proper education so that they may ward off their superstitions and ignorance. Man-killing should be treated as an ecological reaction to the gross over-exploitation of natural resources carried out by the humans for years together and the answer is also to be found by the maintenance of a delicate but complex balance in nature only. There is no other easier alternative.

During the decades 80s and 90s several field Directors have done commendable job. Mention may be made of the names of Pranabesh Sanyal IFS, Vinode Rishi, IFS and P. Sengupta, IFS and others. Sanyal prepared the Management plan of the Tiger Reserve. Under the guidance of all these officers a good number of sweet water ponds were dug and the use of masks/fibre glass shields were set in use by the permit holders. They also experimented placing of human dummy, fitted with electrical devices to scare away the man eaters from the human beings, obtaining only partial success in all these trials.

Sanyal carried out intensive study of the food habits of the man-eaters. Another achievement was bringing the permit holders under cover of Insurance schemes for undertaking hazardous profession and arrangement for payment of compensation money to the next of kins of death victims and to the injured persons.

MORE ON TIGER AS VIEWED BY SY MONTGOMERY

(Source: Spell of Tiger, 1991)

"Wherever it is found, the tiger evokes a reverence, dread and wonder accorded to no other animal".

"Everywhere it roams, the tiger is credited with powers beyond those expected of any wordly animal. Nowhere are those powers morefearful than in Sundarbans, largest delta in the world. Here unlike any other place on earth, tiger regularly hunts people, hundreds die every year in the tiger's jaws."

"Here the tiger is feared but not hated, here it is worshipped, but not loved, for here tiger is a sacred creature who rules an enchanted land."

"In this book the tiger will carry the readers on a spiritual journey to a land where nature and God are one. But we may soon be unable to make, for soon there may be no more tigers to carry us there."

"The tiger must accomplish it drama. It must hunt, it must eat, it must raise its young. And if the Tiger is the guardian of Sundarbans forests, it is because it kills and eats people."

"The tiger's power permeates Sundarbans just as salt flavours its rivers. Most of the tiger does not choose to be seen, just as you cannot see the wind, but you can see what the wind and the tiger have touched."

Sy Montgomery did not fail to mention about the village Amarpur near Gosaba. This village has been designated as 'Vidhaba Palli' or 'widow village' as in each of its 125 families has a woman whose husband or brother or son was killed by a tiger. She mentioned that about the end of 19th century 4218 people were killed by tigers in Sundarbans.

Human fatalities in Sundarbans:

- 1018 nos. of 28 years i.e. 32.8 persons/year

Honey Collector	–	213 nos.
Fisherman	–	531 nos.
Timber collector	–	45 nos.
Others	–	229 nos.
Total number of tigers in 1999	–	284 nos.

A UNIQUE PERCEPTION ABOUT MAN-EATING TIGER

No animal on earth has been integrated as a part and parcel of human-life style and influenced their Super-animal conception anywhere in the world to such an extent as Sundarbans Tiger. This animal has been unique in shaping social ecology of the people of the region. The inhospitable terrain (mud, tide, salinity, immumerable protruding plant roots, cyclone, storm, crocodile, snakes etc.) has moulded the animal with majestic characteristics, that have branded the animal as most cunning, strong and having supernatural powers, thousand of legends depict the elusive vile character of this magnificent animal that has caused fatality in the lives of poor wood-cutter, honey-collector and fisherman of the terrain, who eke their livelihood in this hostile terrain only. They die in large numbers, yet they come to earn their livelihood without bearing any grudge against the animal, which is unique and a bewildering example of man-animal interaction.

Gunins recite mantras in which they call Banabibi as 'Ma' and seeks her protection, the way she protected them in their mother's wombs. The Gunins address 'Dakshin Ray' as father, saying that they were entering his domain (the forests) to earn their living and beg to spare them from his wrath.

Some Gunins chant the mantras as follows:

"you dwell in the forest,
you, deity of the forest,
I am your humble son.
Leave me, I pray, and leave my men in the area,
and go to some other place."
and also
"O Mother
Thou who live in the forest.
Thou, the very incarnation of the forest.
I am the meanest son of the forest.
I am the meanest son of yours,
I am totally ignorant.
Mother! do not leave
Mother! You kept me safe inside your womb
for ten months and ten days.
Mother! Replace me there again,
O Mother pay heed to my words."

There is a saying in Sundarbans – if someone gets a disease, there is a doctor for him, if someone is attacked by a ghost, there is a Gunin for him. But if someone is caught by the tiger, there is no resort.

"Even the best Gunin cannot protect a guilty man", Adhir Krishna Mridha says "All my life I have been struggling against sadness." Adhir's (52) father, brother and son were killed by tigers. His wife was eatern by a crocodile. His daughter drowned in the river. His house was struck by lighting and was burned to ashes.

Adhir turned to be a Gunin. In a dream two boys appeared before him and uttered seven words, "Whenever you go into the forest utter these seven words faithfully" the boys said, "if you are pure and those with you are pure, you will be safe, you can forget your fear". Those seven words were thought to be fool-proof. But the very year Adhir entered the forests with five boats to collect firewood and a man was taken away by the tiger.

Mantra, that keep crocodiles away:

"Dust from heaven!
Male and female sharks,
Male and female crocodiles, dangerous
Ones with sharp teeth,
If you come to bite or attack,
I charge you;
Stay away for day and night,
This mantra, like thunder is a weapon in my hand,
So I charge you in your mouth, in your teeth,
To stay away for day and night."

Straying and use of tranquiliser gun (Ketasete medicine)

(Only a few incidences quoted)

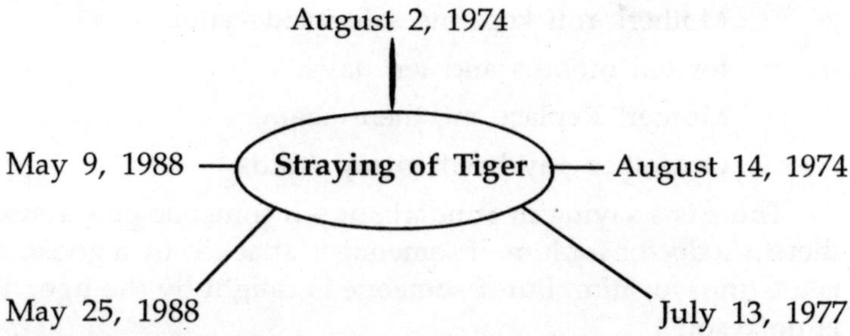

The above chart indicates very few examples only. There are several such incidences which are not enumerated here.

SY MONTGOMERY OBSERVES (1991)

"Oodly, tigers seldom stray into the villages. When they do, they do not stay long and they do not hunt people. But in the forest they consider human beings to be suitable prey.

No one keeps track of actual number of people killed. Unofficial figures indicate the number of people killed to be about 150 per year."

WHY SUNDARBANS TIGERS STRAY IN VILLAGES

Incidence of Sundarban tigers strating in the villages, adjacent to forest was never heard prior to 1974. Since 1974 till 2003 there have been a good number of such incidences. On in depth analysis it was found that the tigers preyed on domestic animals only. There was perhaps no tiger or man-eaters preying on human being. Forest officers used tranquilliser guns to capture them and rehabilitate them into the forest.

SOME ADDITIONAL INFORMATION ON SUNDARBAN TIGER

On the whole such operations revealed that forest offices were very successful in the art of capturing tigers and no casualty of tigers, so tranquilised ever occurred.

But reasons for such straying must be ascertained as to why do they stray out of the forests:

- Starving tigers not getting adequate prey animals in the forest may stray out.
- Inability of the aged tigers to prey upon the prey animals in the forest area may stray out.
- Vigilance and alertness on the part of the field staff of forest directorate outwitting the man-eaters inside the forest from over powering them may be another reason.

From the facts and figures to hand it can be inferred that the number of man-eaters are small as the straying tigers, which are apparently non-man-eaters, did not attack human beings.

During 1990s and there after, 1 to 2 tiger strayed into villages every year which were successfully captured by using tranquiliser gun and rehabilitated in the forest.

The villagers deserve praise as they have been conscious not to kill tiger so strayed; they helped the forest staff in their effort in capturing tigers.

So the question still remains unsolved as to what per centage of the tiger in Sundarbans could be classed as man-eaters?

Man-eaters possess extra-ordinary intelligence to comouflage and protect themselves from human beings:

- They remain in continuous movement not to be leave pug-marks near human habitations.
- Pucking, prusten and parring are various sounds uttered by tiger at various situations.

The air of Sundarbans is pervaded with hundreds of years of innumerable poors' heart rendering tales of death of near and dear ones in man-eaters' jaws, yet the terrain has attracted the people, who took the sacrifices, as granted as the acts of nature and the denizens were innocent and the death was the act of Tiger-God, that reigned the terrain.

A SPECIAL BEHAVIOURAL PATTERN OF TIGER

(Classification of Sundarbans Tiger)

Percentage of typewise population of tigers in brackets etc.

The chart depicts a modified behavioural assessment, that the author conceived in early 1970s, when field study used to be based upon a limited number of information (Ref. P-91 of the author's publication, entitled Sundarbans Mangrove (1989). However, it is just an assessment.

Tiger experts assign the reason for turning of some tigers to man-eaters as follows:

- Physical weakness, old age and apathy for difficult adventure.
- Dearth of prey animals.
- Casual testing of human flesh and blood, attracting them (tiger) to have more of such food.
- Physical disability due to some severe wounds, that restrict them to chase fast moving prey animals in the forest.

In Sundarbans conditions are different where some per centage of tiger are considered to be inherent man-eaters and some notoriously ferocious.

The reasons, assigned for some tigers being inherent man-eaters are:

- Extremely difficult terrain with muddy soil, tidal inundation and dense and thick vegetation.

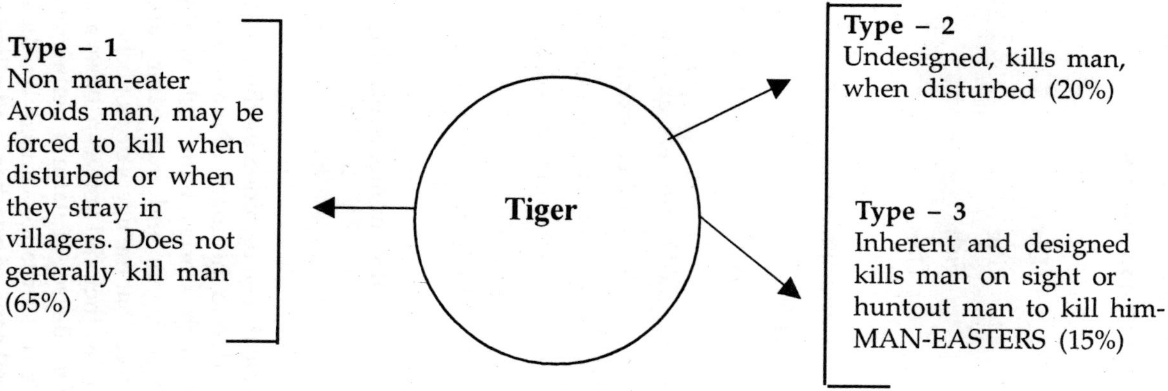
Type – 1
Non man-eater
Avoids man, may be
forced to kill when
disturbed or when
they stray in
villagers. Does not
generally kill man
(65%)
Tiger
Type – 2
Undesigned, kills man,
when disturbed (20%)
Type – 3
Inherent and designed
kills man on sight or
huntout man to kill him-
MAN-EASTERS (15%)

- Dearth of drinking water. It is thought that drinking of saline water over generations have damaged tigers' kidneys and livers, causing acute physical irritation.

FOOD HABIT OF TIGERS OF SUNDARBANS

Tiger's Food

Human Flesh and blood	*Deer
Crab	*Pig
Oyster	Fish
Birds	Honey
Reptiles (Varanus Sp.)	Monkey

(*) marked comprise principal and bulk food

ABOUT THE TIGER SUB SPECIES

To day 5 of 8 sub-species are surviving. Some facts are as follows:

- Indo-Chinese – 2000 numbers existing
- Sumatran - 650 numbers existing
- Siberian - 150-200 numbers existing
- Chinese – 30-60 numbers existing
- Bali - Extinct
- Caspian - Extinct
- Javan - Extinct
- Indian - 3186 numbers existing

TIGER-MAN INTERACTIONS

(Analysis of true incidents)

Hundreds of true incidents have been recorded by the author on human fatalities and injuries due to the attack of man-eaters; he has, however, selected 21 such stories which show the behaviour pattern of Sundarbans tigers and man-eaters and human interactions. The author feels that the tigers do not behave, now, (in 2005) in the similar way as they did between 1950s and 1960s because of forest depletion, dearth of natural herbivores, poaching, trapping of animals, disturbance caused by illicit fuelwood collections, international gang of poachers and pilferers and heavy biotic pressure of poor human fringe population and indirect pressure of populus city of Kolkata.

The study generated data from field investigations were scanned, analysed and re-written by the author – A.B. Chaudhuri, which were given a fine touch by two eminent persons – P.K. Sarkar, IAS. (Retd.) and Ex-major S. Roy. All these were read out and recorded in the proceedings of International Symposium on Project Tiger, held in New Delhi in 1989 (Feb 22–25).

True Tiger Stories

The stories narrate exciting encounters with the tigers in the sinister beauty of the wild, swift flowing water, armoured deadlines of the crocodiles gliding noiselessly through water in quest of prey, mating call of the tigress and answering roar of the mate prowling silently along deserted jungle paths. Pathetic stories of their struggle for survival against a powerful and cunning beast which is still considered mysterious. Yet these men have observed meticulous silence about their ruthless extermination: rather they wanted to coexist. Whatever be the scientist's view on the overlay of superstition, the stories reveal certain interesting insights into the preying habits of the Sundarbans Tigers. These are summarized as follows:

Behaviour pattern no. A.

True Story 1

They are superb, long-distance swimmers; they cross rivers at right angles to the current once the crossing is decided, it is undertaken with a firm determination; and boat coming in the way is attacked; in the water they are just as strong as on land; and extremely crafty fighter with indefatigable energy.

Mayadwip River, April – Afternoon – 1969

A Motor launch was gliding smoothly into the calm; glass-smooth and crystal clear waters of the mighty Mayadwip river. It was the season for honey collection and the entire atmosphere was surcharged with the sweet fragrance of Khalsi (Aegicerus majus) flowers with numerous swarms of rock bees busy collecting honey. But the keen eyes in the launch were riveted on a black, moving object in the middle of the river.

The launch was powerful with a steel hull and manned by a crew of ten. The serang (the pilot) steered the Launch towards the moving object which resolved itself as a full-grown tiger, crossing the river. Shikar stories about the Sundarban tiger infuse these animals with immunity in water from fatal shots as only their nostrils peep above the surface even in rough weather. A strong wind had picked up in the meantime and the Launch started violently rocking on the crests of the waves. Attempts were made to shoot the animal. Five shots were fired in quick succession from the unsteady launch. But the rocking diverted the bullets well away from the animal which was closely watching the proceedings.

The crew set in, aiming lumps of coal at the swimmer which casually blocked the few accurate missiles by its broad left paw. Long bamboo poles, the normal equipment of launch, were thrust at the tiger but the ends were gripped and it caught hold of the pointed end with both paws and chewed it up until reduced to shreds. The exhausted crew were getting more and more nervous at its ingenuity.

The serang repeatedly rammed and over-ran the animal to suffocate and drown it, criss crossing the launch over its head. The cook splashed hot water on the animal. But neither any injury nor exhaustion was noticed. This charmed existence of the tiger caused a section of the superstitious crew to bolt themselves in a cabin below.

The serang was perplexed but the animal temporarily solved the problem, by getting air borne, raising its body in mid-air and effecting a smooth landing on the small dinghy attached to the launch. Having climbed the dinghy, it lay its head below the wooden planks to register its tiredness and fear. Fear, in turn struck the crew line an electric shock with the thought that the dinghy would now be used as platform to climb on to the launch. To stave off such an eventuality, the two rope was extended considerably with the help of another rope.

The battle of nerves was reaching it climax. The panic-striken crew were petrified. At long last it was decided to drag the dinghy with the tiger to the timber coupe of Herobhanga where the animal could be brought under control with the help of the coupe staff in possession of rifles. It was a long way and Matla, a mightier river, had to be crossed. As the launch entered Matla, its rushing waves overturned the dinghy and the tiger was tossed back into the water. It swam ashore completely fagged out, looked at the helpless and curious crew of the launch and leisurely walked away turning its head at the launch several times before vanishing into the mangrove forests; It had fought a grave battle single handed for three hours against man and machine and had won.

Behavioural Pattern No. B

True Story – No. 2

They appear all of a sudden in the most unexpected places; They attack in broad daylight, aiming mostly at the most vulnerable point - the right side of the prey's neck so as to pierce the jugular vein and cerebellum.

Chotohardi Block, Time 4.30 P. M. 28th December, 1971

It was at the junction of Nekrikhali and chotohardi rivers that Kirpa Mandal, a permit holder, was busy catching crabs with his associates. They had a good catch. But the high tide was just beginning and they decided to continue for another hour before dropping anchor for the night, blissfully unaware that they were being watched and stalked for the past several hours by a notorious full-grown man-eater of Chotohardi.

Obviously of the dangerous surroundings, Mandal was concentrating on his work, sub-consciously reassured by the thought that his boat was in the mid-stream on the wide expanse of water. In their pre-occupation, however, the fishermen had overlooked the fact that the channel of Nekrikhali, further ahead, narrowed down to a bare ten metres rendering the boat vulnerable to a tiger's leap from the bank; and they would be passing the narrow neck in another forty minutes. The intelligent tiger, however, had anticipated the course of Mandal's boat and had silently marched off in advance to take up his position behind a Keora tree (Sonneratia apetala) along the bank of the river at its narrowest. One of the fishermen at this point instinctively scanned the bank, and with hawk-eyed accuracy of vision, spotted the tiger. With characteristic courage, the fishermen stopped the boat and attempted to scare away the animal by shouting with their powerful lungs. The undaunted tiger came out into the open and pondered his next move. But the shouting brought a

Forest Patrol boat to the scene and the disappointed animal turned back into the thick mangrove with a snort of disgust.

True Story No. 3

Khatuajhuri Block: 5 Time 5 P.M. December, 1949

Mohan Ganguly lowered his hookah and gave a lop-sided smile. A timber contractor by profession, he was narrating how a reign of terror had engulfed Khatuajhuri. One after another the timber workers fell to the wily man-eaters, although this block had never been considered dangerous. All precautionary measures proved ineffective.

Forest Department staff were stationed with rifles. But the animal took its ninth victim. Ganguly was shocked. Of the nine killed, only three bodies could be recovered. As the body of the last victim was being carried back by a party of camp-mates, the feared cry 'tiger, tiger' echoed. A giant of a tiger appeared ahead of the party, the slanting rays of the setting sun making its dark tan shine against its black stripes. The time was about 5 p.m. The dead body was hastily dropped as the entire party ran for life and dived into the river not realizing that the water was still more dangerous owing to unseen whirlpools and multi-directional currents, not to speak of sharks and crocodiles. It was all over in a matter of seconds.

Ganguly's vivid description, was a far-away look his eyes, almost re-enacted the horrifying and pathetic scene with the frightened swimmers helplessly floating downstream and the man-eater leisurely following them along the bank of the river within easy sight. Death was the closest follower with the twin weapons of drowning by exhaustion or being caught alive by crocodiles, sharks or the tiger. Every second seemed to be taking these lost souls closer to doom. But it had been decreed otherwise. The crew of a "Sundarbans Despatch Service" vessel bound for erstwhile East Pakistan sighted the tragic scene. With the help of a speed boat, they rescued the unfortunate persons took them to the camp. From a distance the tiger watched the entire proceedings closely accepting his defeat but only temporarily.

Forest Department authorities ordered a temporary suspension of all work. But necessity knew no law. At the request of the permit holders the work was allowed to be resumed.

Although security arrangements were tightened considerably, two more helpless workers were carried away by tigers in quick succession. Finally, with the sacrifice of eleven persons at altar of Dakshin Ray, the Forest Department officially closed the camp that had claimed so many lives. Mohan Ganguly fell silent, his hookah forgotten.

True Story No. 4

Chamta – 6, Time – 3.30 P.M. August 1965

Gosaba – 6, Time – 3.30 P.M. August 1965

Nurul gazi had passed his 65th year in a happy home. His source of income was collecting Golpatta (Nipa fruiticans) for roofing and he was one of the oldest permit holders with a vast knowledge of the terrain and its denizens. Caution was his second nature and on this particular day he had posted two men to watch and warn, while he himself concentrated on his work in Chamta–6.

Lunch was over. About 3.30 p.m. Gazi noticed the absence of one of the watchers and approaching the spot where the man was posted, discovered the man-eater's pugmarks on the receptive soil. A man of determination, Gazi followed the track with the other watcher till he sighted the man's dead body. Keeping his distance, he began scanning the area. The next moment, the man-eater seemed to float silently through the air from behind and had the right side of the neck of Ashraf, the 40-year old second watcher in his jaws. Though Gazi was able to scare away the animal, his only reward was to be able to carry, along with others, both the dead bodies back to waiting boat.

Seven sad days had passed and Gazi was forced to shed his sorrow and to start his work once again as money was running low. He landed in Gosaba-3 with four companions and was horrified to be confronted with fresh pugmarks of the dreaded tiger. To add to their miseries, a discussion on the plans for work and watch was interrupted rudely by the tiger's roar close at hand and all four of Gazi's companions were frightened into unconsciousness. This time luck was with him as he finally succeeded in scaring away the animal before it could launch an attack. Fortune was in to him for a change.

True Story No. 5

Panchamukhani Block – 5, Time 9 A.M. 29th November, 1971.

Harihar Samaddar with his friends, Nibaran Mandal and two others got off their boat to collect dry twigs for firewood and to check the spoors of different animals. Harihar busied himself with noting down the pugmarks and their sizes. Having found some dry twigs Nibaran collected some and carried them to the boat enchored hardly 500 metres away. On his way back from the boat, Nibaran was half-way when he detected fresh pugmarks of a tiger. Frightened movement of monkeys on the trees confirmed his fear and nervousness made him shout out to his companions. Before any help arrived, the cunning and stealthy man-eater caught him and was about to carry him away when Harihar's crackers started bursting and the two other men came to Nibaran's rescue. His lifeless body was abandoned by the tiger which disappeared into the jungle.

On examination it was found that the lugujar vein and cerebellum had been pierced and the spinal cord at neck was broken and Nibaran had died instantaneously.

Harihar has not, yet reconciled himself to the incident which continues to baffle him with its lightening speed and unexpectedness.

True Story No. 6

Panchamukhani – 5, Time – 10 P.M. – August, 1970.

Satish Mandal at 50 had been in the timber trade for 30 years and had lost many of his friends and relations in the Sundarbans.

He was camping in Panchamukhani. It was 10 P.M. and as he was about to turn in for the night, he jerked upright as he felt slight jolt of the boat. It struck, him as somewhat unusual. He gripped a stick, grabbed a torch, and came out on the deck of the boat. The pencil of rays form the torch revealed a full-grown tiger struggling to climb on to the stern of the boat. The boat was big one and the stern was quite high above the water and there was no dinghy to use as platform. These factors were frustrating the tiger's attempts to jump on board. Mandal summed up the situation immediately and woke up his men.

They lustily shouted in unison and struck the tiger with strong Goran (Ceriops spp.) sticks. The tiger swam back to the shore, leaving the entire party unharmed.

Behavioural Pattern No. C

The tiger climbs on to a boat anchored in midstream without a sound; not even a splash is heard, the boat does not rock and the silence is unbroken; the tiger then selects its prey; the boat rocks this time as the tiger jumps off the boat into the water with the man securely clamped in its jaw for a quick exit; it is then too late to intercept it and save its prey.

Salnikhali Patrol Party Staff, Time – 11 P.M. August, 1966.

As with smugglers in the Sundarbans, tigers too avoid Government boasts and patrolling forest staff.

A forester happened to step outside the cabin of a boat around 11 P.M. one night. As he settled down on his bed after re-entering the cabin a soft thud was barely heard from the stern that jolted the boat slightly. All the inmates of the boat were cautioned. To ascertain the cause, a guard moved out with a torch. He scanned the front end of the boat and then turned the light towards the stern where he faced a tiger in a corner, ready to jump on him. The guard although scared had enough presence of mind to jump back into the cabin and wake up the inmates. The disappointed tiger had to jump back into water and swim ashore as his manoevure failed.

True Story No. 7

Bagmara – 8. Time 11 P.M. May, 1969.

Sanyasi Mandal was saying: "I opened my eyes from deep slumber. It was about 11 PM. A pair of eyes glowing in the poor, flickering light of the kerosene lamp in the boat and a warm putrid breath was fanning my face. The animal stood among the four of us while choosing its prey. I tried to shout but my voice failed. I have seen many tigers but nothing like the one I saw this time."

Sanyasi was in the fish trade for a long time and knew the terrain very well. A bright young man of 25, with thick tufts of curly hair and a complexion like varnished ebony, he often went in parties of five or six for fishing in the creeks ends streams meandering through over-hanging Goran and Garjan bushes. Each fisherman could hardly earn anything more than a meager sum of Rs. 2/- to 3/- a day while the middle man raked the lion's share of the profit. A lovable young man, Sanyasi kept close contact with the staff of the Forest Department. He was aware of the multidirectional currents of the channel running east-west and other hazards of the terrain.

It was customary for the fishing-folk to get under the bamboo canopy of the boat after the day's work. On this day too, they had sat there and chatted for hours before turning in. Sanyasi cannot recall why he suddenly woke up to find the man-eater towering above him and his team mates right under the canopy. The animal selected Ananta Mandal, a man of 40, lifted him up his jaws and jumping out of the boat, vanished in the darkness. Nothing could be done as it all seemed to be over in a second.

Sannyasi recollected his promise to Ananta's wife that they were going together and would return together. He was disturbed over the thought of having to face the deceased's wife. He continued to say that tigers, crocodiles, sharks and venomous snakes had to be encountered daily by them in a hostile terrain, boasts sink in storms, dacoits rob them of their belongings, drinking water and edibles; yet they continue their professional work ignoring the sorrowful tears of wives, mothers and children.

He attributed supernatural powers to the tiger; otherwise how could they get inside the boat so silently without an indication of their presence.

Behavioural Pattern No. D

When challenged animals pause and stand still, when they scent any danger, they bolt. Are they cowards? Why do they attack from the rear and to the right or mount a boat like a thief?

True Story No. 8

Gosaba Block. Time–9 P.M. July, 1968.

Rashik Mandal narrated his thrilling encounters with tigers; he had escaped unscathed, while his team-mates fell one after another. He said, man-eaters were everywhere and they migrated from block to block. He considered the tigers of Chamta as the most cunning and notorious.

He had once had the experience of a tiger boarding his boat. Fortunately that night all the inmates were relaxing but had not gone to sleep. The boat tilted just a little and unless one was experienced it was impossible to associate the tilt with the presence of a tiger on board. However, that night the tiger was scared away and everybody was saved.

True Story No. 9

Chotahardi Block Time-10 AM. April, 1970.

Bishnu Mandal, aged 60 was a honey collector. He and his guru (preceptor), both priests, were in a group. The guru had a premonition of grave danger and decided that the men should stay together for sometime. His hunch paid off and they sighted a huge tiger as they were detailing their duties. It was a Friday and they spent the rest of the day in intensive prayer.

Early on Saturday morning they stepped out on the dry bank of a creek. Blessings were invoked form the Tiger-God. Mantras (incarnations) were chanted to keep the man-eaters away and also to mobilise them. In the process, Mandal lagged behind the rest of the party. A huge tiger appeared as if from nowhere. But Mandal's courage prevailed and he faced the animal and shouted as it:

"You rascal where do you come from?" Mandal had a sharp cutting implement with him. But the tiger was hungry, snarling, twitching its tail with its facial muscles expanded, spitting frequently and the saliva oozing out, hind legs grazing the ground; it visibly grew to double its size. Though death was imminent, Mandal did not give into panic but held the sharp instrument high, ready to strike. This caused the animal to pause. Mandal's eyes locked with those of the tiger while he was backing slowly. He touched the ground with both hands and chanted a mantra. The animal scented danger and suddenly retreated. Mandal was rescued and found to be running high temperature. His eyes were "blood red" and it has remained so till today. A few minutes strain had changed his facial expression altogether.

True Story No. 10

Chamta Block, Time-9 A.M. April, 1967.

Golam Muhammed of Gosaba, had been collecting honey for 20 years. As an eye-witness, he described how, under pathetic circumstances, he had lost seven of his close colleagues on different occasions.

But the present incident which he was going to relate was different. It was a bright and clear morning. A mild breeze broke the gentle swell of the river into dancing little waves. Reclining at ease on the prow of the boat Gulam Muhammed was watching the golden brown colour of golpatta and the emerald green of the Dhani grass (*Oryza coarctata*). The tide was running and the boat stood close to the bank. Golam, along with some others, was relaxing as he expected some of his men to join him. His son, cooking inside the boat, suddenly shouted "Tiger, Tiger" and the animal was over Golam. With the impact of the jump the boat moved and instead of bringing Golam's neck between the canines the animal could only put the front paws on his knee. Golam, out of sheer fright and nervousness, caught hold of the man-eater's broad head and shouted in a frenzy: "You rascal, get away; don't you know where from you have come?" almost simultaneously his other colleagues arrived; repeated shouting and beating with the sticks made the tiger retreat into the forest. Golam showed his scar which was still giving him trouble. He, however, proudly asserted that the man-eater had to depart as he had chanted mantras.

Behavioural Pattern No. E

Men die; they lose their near and dear ones to the tiger, every time they go back home. They decide not to come out again yet they come in this saline retract soft feelings and fear vanish; It makes a man ruthless, tough and teaches him to fight for survival.

True Story No. 11

Block Gosaba-2, Time-4 O.M. April, 1970.

Gajan Howley, a honey collector of Amtoli, had been on the job for 28 years. He had heard and seen many people die. He, like others, had accepted it all as inevitable.

This particular afternoon Gajan selected a creek for his collection as a swarm of honey bees was detected in the sky. Four of them got off the boat and proceeded deep into the forest in a row each hardly 20 meters apart form the next. Gajan was talking as he proceeded. After sometime he looked for his companions but none could be seen. He looked carefully again and found that a man-eater which had so long concealed itself and seized Pratap Howley by the throat and was holding on till all signs of life vanished; while two others lay unconscious on the ground at the sight of the man-eater as it attacked Pratap. Intense love for his friend prevailed over fear and nervousness. Howley held a Goran stick in hand and chased the tiger as he shouted. At one stage Pratap's leg - he was dead by then - got entangled in a Goran bush and the tiger could not extricate it. Gajan caught hold of a leg. A tug of war began - the tiger pulling the body from the front and Gajan pulling one leg from behind. In the meantime two of his men regained consciousness and joined Gajan in beating the tiger with Goran sticks. In desperation the tiger finally left the dead body and vanished into the forest.

True Story No. 12

Gosaba Block–3, Time–4 P.M. April, 1963.

A party of honey-collectors had just returned to the boat with lengthening of shadows when a swarm of low flying bees attracted Ananta Mandal's attention. He thought a few honey-combs must be located close and set out for the combs with two other collectors.

To ascertain the exact location Ananta climbed a tree. A comb was located and two of them proceeded towards it.

The area was bushy and almost impenetrable. As he claimed down Ananta saw a tiger crouching stealthily towards a companion of his from a very close distance. Fright paralysed him and he could neither shout nor move. He saw the animal twitching its tail and just about to leap on its prey. The unfortunate man hardly knew that his death warrant had been signed.

As the animal swiftly leaped high in the air, the victim had some premonition of danger and turned twice so that the tiger missed the target by inches - a bush having intercepted. The scene that followed for the next few seconds was unique for the Sundarbans. The tiger was trying to get a grip with his jaws while the three honey collectors were lurdly beating the ground with branches and shouting lustily. This was perhaps a danger signal for the tiger and it vanished hurriedly.

On being interrogated, Mandal stated that they all were aware of the danger and the risk involved but these wee inevitable. In his opinion this tiger was perhaps a young and inexperienced one.

To eke out a living, they have to trek such risky and difficult, mud deep terrain where venomous snakes, deadly crocodiles and man-eaters are encountered frequently.

Behavioural Pattern No. F

They are invisible, they appear to be supernatural, they approach with supreme stealth even in broad day light; they attack on the right side of the neck, piercing the jugular vein and the cerebellum.

True Story No. 13

Chamta Block, Time-9 A.M. November, 1970

Maherali Mallick, 60, worked in a Golpatta coupe. He had terrifying experience of a life and death struggle of the permit holders for survival. In early 70 he had seen two of his men being taken away by man-eaters.

While seven of his men were engaged in working the coupe, two were specially employed as watchers to guard the two vulnerable flanks. It was about 9 A.M., work having started at 8 A.M., when some of the men noticed that one of the watchers was missing: a search revealed the fresh pugmarks of a tiger and a few fresh drops of blood.

A search-party was arranged with a 'Priest' leading it. They had trekked about 5 kilometers tracking the blood drops and the body was found in a Hental (Phoenix) bush. On the 'Priest' lay the responsibility of recovering the dead body. On careful search it was found that the tiger was not in the vicinity.

As the priest went close to the dead, the man-eater roared savagely and pounded on him from the right and, in full view of the other men, carried him away as if to mock at the efficacy of mantras. It was a pitiable story of the death of two hapless men.

True Story No. 14

Chamta Block. Time – 7 P.M. May, 1969.

Another honey collector, Nagar Ali of Katakhali, considers the tiger a coward. He said they act like thieves, coming stealthily and attacking from behind.

He cites the incident when he and his team-mates were on a boat anchored in Chamta river. It was dark; food was being cooked and everyone was relaxing, some were singing and some were engaged in conversation. Malek Molla was seated in a corner and was not participating in any discussion.

They all became alert as a sound came of something heavy splashing in the water. Where was Molla? It certainly seemed that he had vanished. Torches were flashed all round and the faint figure of the man-eater with 'Mollah' clamped between its canines, was spotted on the far bank.

Molla's half-eaten body could be recovered only the following day. The Sundarban man-eater eats the soft stomach first when breaking its fast.

Nagar Ali also recalls that another worker was killed in the same compartment the previous year under similar circumstances.

True Story No. 15

Chota Block-5, Time-8 A.M. February 16, 1970.

Provash Baul, Forest Ranger, was in-charge of a timber coupe near which hundreds of boats, big and small were anchored.

It was a bright, sunny morning in winter, and Baul requested Sunil Mandal (40), a permit holder, to accompany him in his boat. Mandal begged to be excused on grounds of sickness.

Sunil had in his mind a different plan to fish in Khal (channel) as it was ebbtide and the water had receded. Immediately after Baul had left, Sunil took a fishing net and started casting it not far from the anchored boats.

A sudden thud was heard and a tiger vanished with Sunil just like a cat holding a fish.

There was no indication of a tiger so close to the anchored boats where hundreds of workers were camping. Places to hide are very scarce in a working timber coupe. Yet it was there that a life was lost when Sunil vanished for ever as if an invisible force had dragged him.

True Story No. 16

Chamta Khola, Time – 4 P.M. September, 1966.

Basanta Reften (40), of Hinglegunj was a worker in a golpatta coupe and had come face to face with two charging tigers. He survived both the charges but went down for several months thereafter with a high fever.

On the second occasion, six of his colleagues were engaged in cutting leaves. One was posted to guard against any probable attach from crocodiles or tigers. Yet stealing through the shadows a tiger came slinking in and took away Bimal Raften (40) from one corner and the others never knew about it for a full ten minutes. The fact was verified from the pugmarks and blood drops.

All the workers were determined to recover the dead body and were following the blood trail and the spoors. After trekking for about half a kilometer inside the forests, they disturbed the tiger as it was eating the dead. The animal displayed unmistakable signs and dissatisfaction and anger as it snarled and growled. Sibu Mandal, one of the men, dropped down unconscious in sheer fright. Once he regained consciousness, he started shouting "Tiger", "Tiger", "Tiger". Sibu could only be carried to the boat after the tiger had moved away from the kill. He ran high temperature for months with regular fits of bervousness and he kept shouting "Tiger", "Tiger"; in his delirium.

Behavioural Pattern No. G

Baulies, or priests, must accompany each party of permit holders; blessings are invoked from "Tiger God", Banbibi or Narayani Ma; the permit holders have immense faith in such priests, as they volunteer to enter the forest first in the belief that they are capable either of immobilizing the tiger or of mesmerizing them into obeying their orders. Many priests are killed in the process.

True Story No. 17

Jhilla Block-3, Time – 7 A.M. February, 1957.

In was a Golpatta coupe and Anil Mridha and others had a priest in their party.

Soon after setting foot on the land, the priest put both of his hands on the root of a Goran bush and chanted mantras for several minutes. Then he moved in a circle with a cutting implement for sometime muttering all the while.

Thereafter the party started cutting the golpatta leavers within an area demarcated by the priest who specifically kept the adjoining ground out of bounds for the workers.

But curiosity led Anil Mridha to approach close to the boundary. As soon as he removed a few leaves, he found a tiger lying mesmerized. Mridha was mortally afraid and ran for his life shouting "Tiger, Tiger". The priest came forward and expressed his disapproval at the discovery. Fortunately no accident occurred in the coupe as the tiger quietly moved away.

True Story No. 18

Lahiripur Village (Jote land)

Dhirendra Mandal described how some tigers cross over to the villages in search of cattle and dogs; and was emphatic that man-eaters do not enter any village. In this respect the man-eaters of Sundarban differ from the man-eaters elsewhere in India. It is through causing annoyance or irritation that people sometimes get mauled or killed by tigers.

Mandal relates an incident. A big crowd had assembled as news reached Lahiripur village that a gunin (priest) had mesmerized a tiger in a plantation of tall grasses. A few shikaris also came with their firearms but could not do anything as they could not take up any convenient position for shooting. An artificial camouflage was erected and the shikaris positioned themselves inside. The priest permitted them to shoot the tiger on condition that substantial offering would be made to the Tiger-God of the Sundarbans.

The Shikaris shot at the tiger, but the tiger reacted by jumping at them although it missed its target. The wounded tiger finally hide inside a small bush across a channel. Once again the shikaris chased it and the tiger returned to the plantation.

The priest wanted to mesmerize it once again but locals disapproved of his doing so as the tiger was wounded.

The priest, however, said that he would make it easy for the shikaris to shoot by removing the tall grass. His efforts only made the tiger jump on him. He was caught by the neck and died on the spot. Subsequently, the wounded tiger was shot and killed.

They select their prey; age may be a factor; the middle-age group suffers greater casualties than the young and old.

True Story No. 19

Malta Block 2, Time – 8.30, May, 1968

Nibaran Mandal (45), of Chotomollakhali was a professional honey-collector and had seen his friends being taken away by man-eaters in various situations. One such fantastic incident he has narrated.

It occurred near the banks of Havila-Duania river on a bright, sunny morning. Ten honey-collectors clambered down to the bank, stood in a line facing the morning sun and the bauli commenced chanting mantras. The surface of the meandering river was crystal clear and calm and smooth as a sheet of glass. The mantras were chanted and the Tiger-God was being invoked.

Suddenly a mild thud broke the silence and Deben Mandal standing in the middle, was seen in the jaws of a man-eater; the sight was akin to a fish in a cat's mouth and he was being taken towards the jungle. Nibaran was next to Deben but due to shock he did not realize what had happened. Four of the collectors however fell down unconscious at the sight. Nibaran quickly arranged a search-party but failed to recover the body.

Results of analysis

Tigers are present in all the fifteen (15) blocks in the Sundarban Tiger Reserve (2,585 sq. km.) and are recorded in all the compartments (65 nos.) of these 15 blocks; in fact 21 blocks of the entire Sundarban forests record the incidence of man-eaters.

- Tigers migrate from block to block as was evident from pugmarks on muddy, sandy and loamy soil and on actual sightings.
- About 40 persons are killed every year on an average (unofficial reports record about 100).
- Maximum casualty is in April and May during the honey collection season, minimum is during the rains. Fishermen are the highest victims; there are some deviations.
- Permit holders of all professions die: in the timber coupes where the maximum number of workers stay and work together (contrary to fishermen who are isolated all over the areas have comparatively minimum fatality).
- Fatality in a particular block do not maintain a uniform level, it fluctuates in a regular rhythmic pattern.
  - (i) Maximum fatality (80%) occurs between 7.00 hours and 9.00 hours, 15.00 hours and 17.00 hours and at 23.00 hours.
  - (ii) The age of persons killed varies mostly between 35 to 45 years in 80% of the cases.
  - (iii) Those killed at night are attacked at their boat after the tiger swam to the boat and mounted on it, selected the victim and jumped with it back into water.

Salinity is suspected to have positive correlation with the ferocity of the animal. Damage of liver and kidney has been suspected (but not confirmed).

The diagram 3 shows that maximum kill is made between March and May. The premonsoon weather mingled with flowering of forest plants and the season of honey collection govern the behaviour pattern of the man-eaters.

Even today it is same after 30 years of this finding. The vulnerable age of people killed by tiger has been found to be between 35-45 (Table 1).

Table 1: Fatalities from tiger attacks among different age classes (1962-1963 to 1971-1972).

Age group	No. of fatalities due to tigers	No. of workers in forests of Sundarbans	% of the fatalities of workers	% of workers in the forests of Sundarbans
16-20	3	1,050	4.2	1.4
21-25	10	2,560	10.8	3.1
26-30	7	5,799	8.2	7.2
31-35	14	7,843	13.8	9.8
36-40	17	6,778	18.2	8.5
41-45	18	9,000	18.8	11.1
46-50	8	13,462	8.7	16.6
51-55	4	13,000	5.3	16.2
56-60	6	14,000	6.8	17.3
61-66	1	5,100	1.2	6.3
66-70	2	1,160	2.7	1.4
71-75	1	908	1.3	1.1
Total	91	80,660	100.00	100.00

Source: Chaudhuri and Chakraborti (1972).

- The tigers understand the human mind meticulously and all their plans of attack are designed on human movement.
- All the forests were classified into types or formation on the basis of tidal level and on their tree associations; the density of such forests were determined and the human casualty pattern in such types and formations enumerated.
- Pugmarks were measured and classified into various size classes and assessment of the younger class was made. It was observed that the younger class was poorly represented.

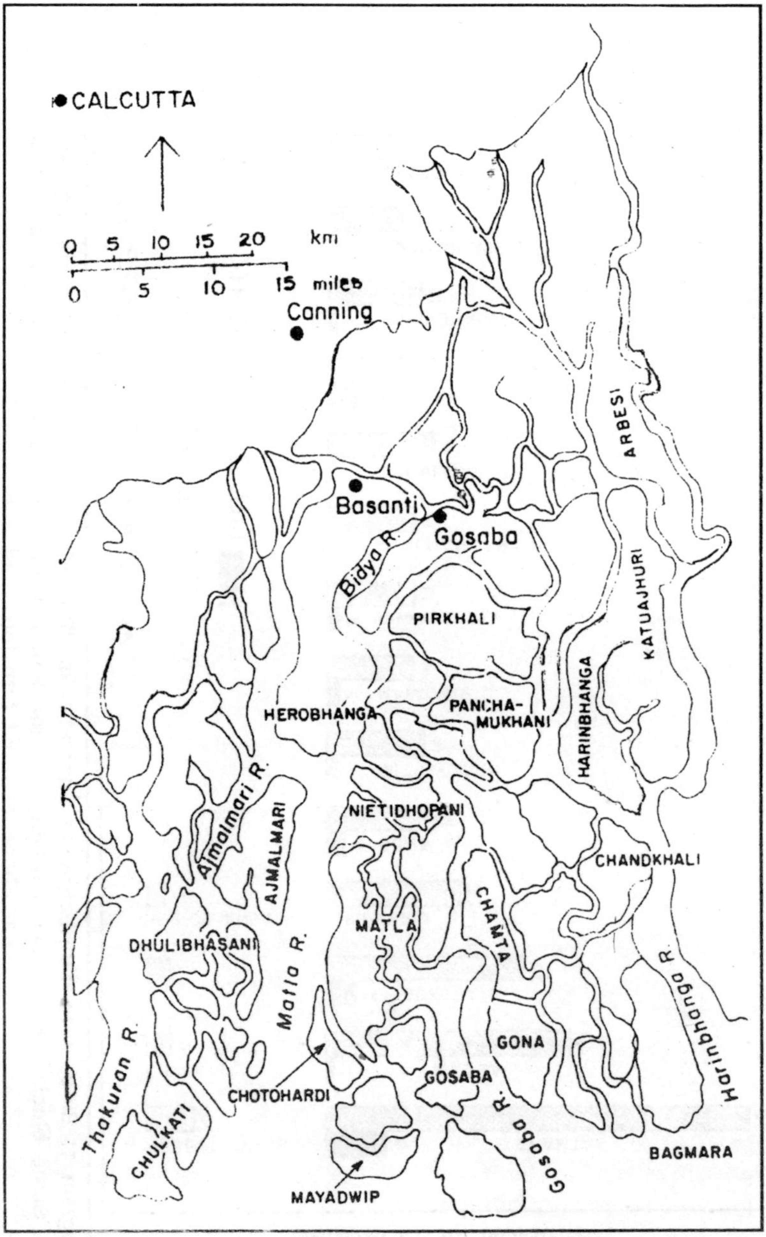

Map - 5 Forest blocks of the Indian Sundarbans

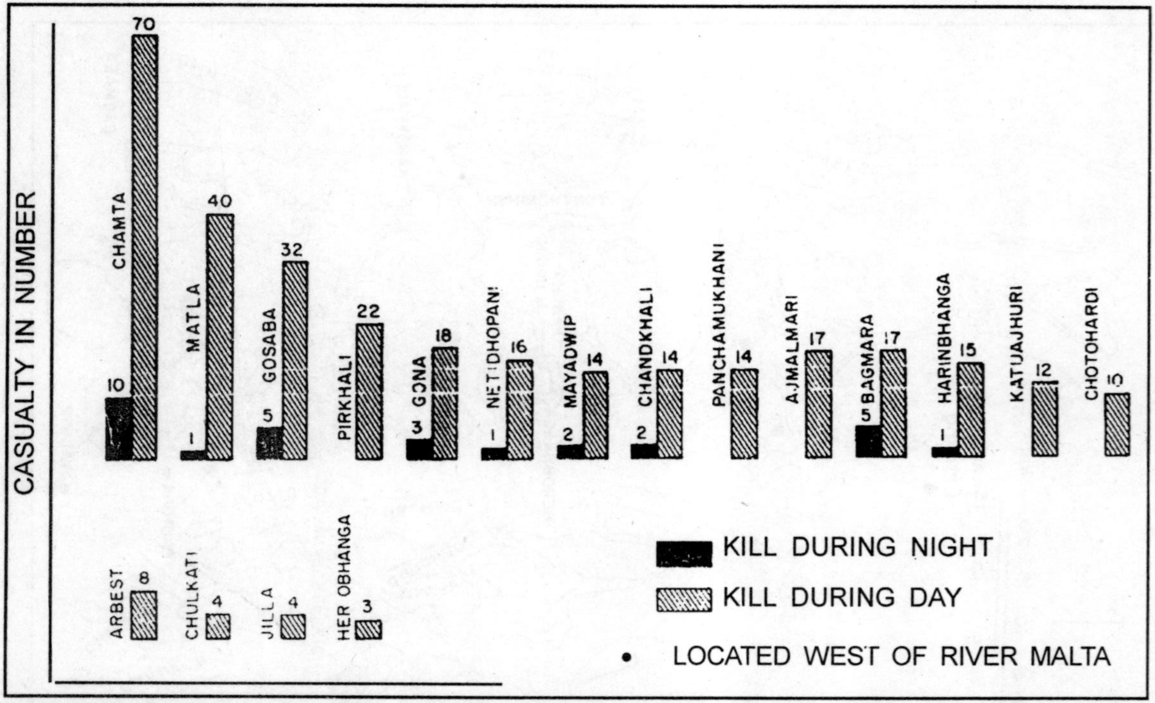

Diagram 1: Block-wise distribution of tiger attacks. Studied in 1970's. The pattern is the same even today. Some blocks cause human fatality due maximum death. The blocks where men are killed at night were abode of most ferocious and cunning tigers.

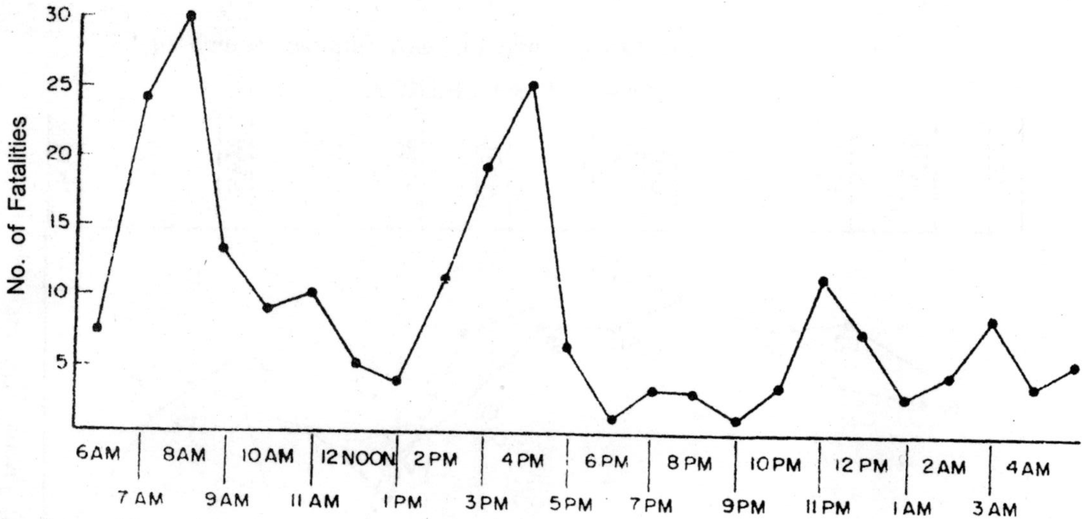

Diagram 2: Pattern of tiger attacks (time) and human fatality. It was established that most kills are made in the morning and in the evening.

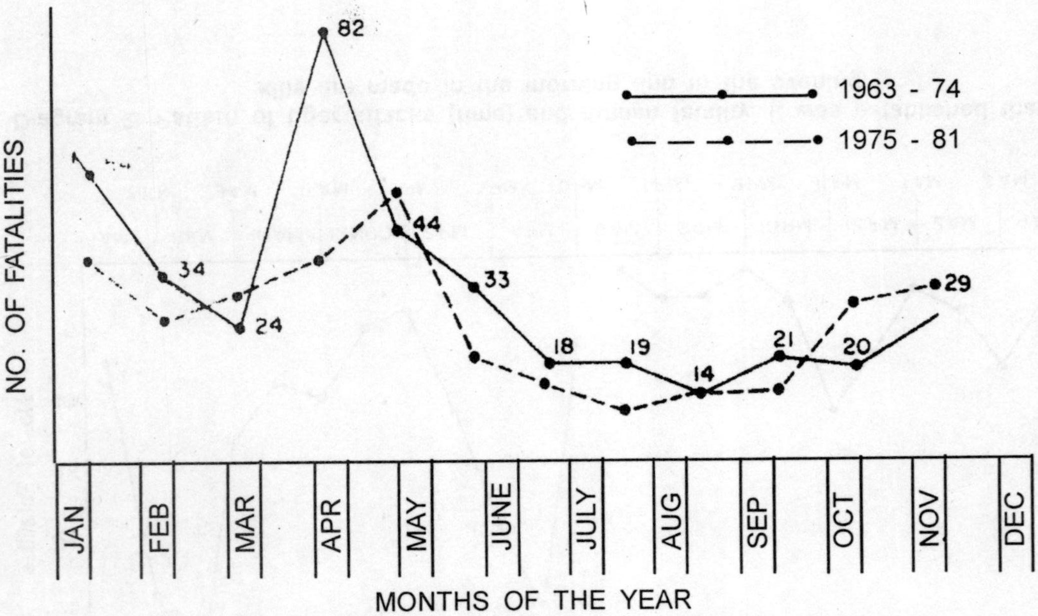

Diagram 3: Monthly trend of tiger attacks and human fatality

- Recovery of the human body could be done only in 28.5% of the cases, whereas in 71.5% cases dead body could not be recovered from the tiger. This indicated the tenacity and ferocity of the man-eaters as the persons engaged in retrieving the dead body were attacked.
- The initial attack was always on the right hind part of the neck and the injury inflicted on the jugular vein and the cerebellum by the right and left paws respectively.
- Pigs and deer were the primary prey animals, but circumstances compel the tigers to monkey, fish, bird, lizard and even crab. They have been observed to eat honey but this may not be a regular feature.
- Striking uniformity – in the number of persons killed in honey collection over the years showed that the number of man-eaters was quite small and static, (on average 17 honey collectors were killed out of about 1300 men engaged in this job each year).
- Maximum fatality zones were in areas of high salinity (but it has not been confirmed).
- The man-eater did not cross into human habitation for predation, which is striking feature of the Sundarbans tiger (It's now a deviation)
- The sex ratio of the female to male was 1 : 3.
- The man-eater dug up recently burned human body to eat it.
- Man-eaters were mostly male. No human female has been killed by a man-eater as per record.
- There was no source of sweet water in the entire tract and although the animals drank only saline water, yet sweet water was an attraction and was preferred to saline water.

Predatory aberration of Sundarbans Tiger must be looked into carefully and steps taken to mitigate the circumstances leading to acquisition of such behaviour.

The genesis and the tragic death of human being due to predatory aberration of estuarine tiger is perhaps due to combined effect of environmental attributes and human activity pattern. The saline water, muddy terrain, recurrent high tide, dense vegetation etc. coupled with difficulty in preying upon animals bring about a psychological change in the tiger behaviour. This type leads to be cyclic and chronic giving the tiger predatory proficiency.

Habitat amelioration is not possible in this tract but the human activity pattern has to be changed. It may be a necessity to isolate the man-eaters.

It follows, therefore, that the study of the habitat and behaviour pattern of Sundarbans tiger is a subject of research and deep study. It is a difficult task. It should be possible to isolate the man-eaters and study their nature. Creation of large enclosures (if at all possible) in the forest habitat and releasing captured man-eaters to study their behaviour should be the subject of present investigation. Investigation on the blood, urine, kidney, liver etc. of this animal is an immediate necessity. It is the research, and not management, ion tiger which is of prime necessity in this estuarine tract. Fantastically high death rate in Chandkahali, Pirkhali-7 and Netidhopani-5 leads one to accept either of the three theories (i) migration of inherent man-eaters or (ii) increase or concentration of tiger population or (iii) man-eating tendency revitalized in the dormant ones showing a rhythmic pattern of activity.

ANALYSIS AND CONCLUSION

The problems involving socio-economic, scientific and administrative aspects of man in relation to tiger of the Sundarbans are specific and unique and attempts to solve them will require considerable time and intensive attention.

The stories describe the miseries, worries and anxieties of the permit-holders who have to encounter man-eaters frequently without any protection, except the blessings of the "Tiger God" and mantras chanted by the priests accompanying them. They die under helpless and pathetic circumstances.

The eye-witness accounts, on analysis, reveal the following features of the tigers of Sundarbans which merit further scientific studies:

(i) exceptional strength

(ii) high capacity for manoeuvring itself

(iii) great power and prowess in swimming
(iv) diabolical understanding of human psychology
(v) favourite hours of attacking human prey
(vi) migratory behaviour
(vii) aggression during full moon and new moon
(viii) adaptation to drinking saline water
(ix) consumption of wide variety of food (subject to research)
(x) preference of middle aged men for prey
(xi) apparent cowardice
(xii) influence of saline water on its physiological and psychological processes (a subject of research)
(xiii) influence of saline water on liver and kidney (a subject of research).

Curiously, the permit-holders' firm convictions that the tiger recognize the boats of forest personnel as the latter carry firearms. This is why the tigers elude them, while the permit-holders come across so many. They seem to be convinced that many tigers were shot, and trap guns laid during British rule. Since the fifties few tigers have been killed. Now pugmarks are visible almost everywhere and incidence of death and injury has risen sharply and is actually quite in excess of official reports.

The author's initial investigational reports on the tiger of the Sundarbans estuary were detailed in a series of articles published in 'Cheetal', 'Science & Culture' and in the "The Bulletin of the Botanical Society of Bengal" during 1972 and 1973.

Besides these and other observations on vegetation and habitat, studies on the breeding habit and bio-ecological aspects on various fish species, honey bee (Apis dorsata) and several species of birds in relation to tigers have been made in a separate series of publications. (*Science & Culture*, Vol. 40, March 1974; Vol. 38, June 1972 and Vol. 39, January 1973; *Cheetal*, Vol. 15, July 1972 and Vol. 15, March 1973). Extensive data have been collected on soil, water, pugmarks, human casualties etc. and those have been analysed. But only a few of these findings fare enumerated in this book. However a few diagrams are presented for perception of the readers. The pattern is same even today.

ANALYSES OF VARIOUS PARAMETERS OF FIELD DATA

Analysis of the above data exhibits that although water salinity tends to increase as one proceeds form the Northern zone to the Central zone and then to the Southern zone of the Reserve, similar pattern was also noticed so far as soil salinity index is concerned, but the frequency of human casualties does not vary according to such zonations. This indicates that the man-eaters of the Sundarbans are concentrated in some habitat formations only, which also incidentally records significantly more salinity. It requires further probe to come to any definite conclusion to ascribe reasons for the man-eating urge of the Sundarbans tigers. It may, however, be concluded for the present that the man-eating urge, a peculiar trait marked in a per centage of the Sundarbans tiger, is governed by cumulative action of physical and chemical environments of the Sundarbans. It has further been noted that the abundance or scarcity of prey animals are not directly related to the human casualty figures to tiger.

The Sundarbans is a natural laboratory. It provides astonishing examples as to how physical and chemical environment can exert influence to the different forms of life. The extent of endogenous and exogenous periodism, osmoregulatory adaptations amongst different organisms were perhaps nowhere so much pronounced and diversified as they are in the Sundarbans, man-eating tigers - astonishing and ferocious - world's largest and most ferocious crocodiles, water and land lizards, crab-eating spotted deer, monkeys and wild pigs, innumerable varieties of crabs and other wildlife, and more than two hundred species of fish, etc., make the area unparalleled.

Further detailed scientific study is likely to throw light into the hitherto unknown biological properties of the Sundarbans.

The Sundarbans tigers are branded as man-eaters since time immemorial, but the entry of government personnel deep inside the forests has not diminished. Men and tigers are living together through ages despite cases of human fatalities. The problem of man-eating in the Sundarbans is thus, nct, something which bewilders the people, as they accept these incidents in their life-style and a living system. The problem of man-eating elsewhere can hardly be compared with the man-eating incidences in this tract. Jim Corbett used to attribute old age, disease and injury to man-eating in the tigers. This does not seem to be very cogent reason of man-eating in the Sundarbans tigers. The reason lies in the fantastic swampy terrain of the Sundarbans with incredible plant density with sharp and pointed breating roots. The physical and chemical environment of the Sundarbans obliterates all limits of land and water animals who are forced to lead amphibious mode of life. In this peculiar eco-system the intruding man becomes the easiest available prey to the adventurous tigers, as he is defenceless in the face of any attack from the tiger. The Sundarbans tigers with diabolical understanding of human nature keep a close vigil on the seemingly dwarf creatures of men while they move in sitting postures in boats and dinghies; Intake of saline water and the soil also is likely to create physiological imbalances not only to the tigers, but also to other wildlife denizens; there are horizontal and vertical stratifications in salinity indexes in the terrain of the Sundarbans and the indexes also maintain seasonal fluctuations. Thus it is not correct to conclude that all wildlife denizens of the Sundarbans should be affected to an equal extent by salinity, as same are likely to be affected more than the others, according to their limits of tolerance to variants and the vagaries of the eco-system. People entering the tigerland of the Sundarbans hardly take any precaution against these man-eating tigers because of their utter callousness, tradition and, above all dire poverty. People go out in open dinghies and boats for months together to eke out a miserable living for them and to support their dependents and fall an easy victim to the circumstancial wily man-eating tigers. This is deemed as a normal fact of life and is sadly accepted as occupational hazard. Payment of adequate compensation in time may also improve the man-tiger interactions to a very great extent. Concerted and all round fringearea development programme may remove the biggest human pollutant, the poverty, in a big way and successful implementation of the same is likely to pay rich dividends in improved man-tiger inter-actions in the terrain.

Some peculiar innovative attempts in electric fencing and battery charger were experimented in the terrain of the Sundarbans sporadically without monitored observations and the protagonists have claimed loud successes from these experiments conducted for the reduction of human fatality. The protagonists, for the time being forgot that man-eating tigers do not restrict their killing operations only to official permit-holders since a large number of persons entering the area without official passport also get killed by the tigers and the success stories do not include such cases who are branded as illegal entrants. Success or failure of an experiment pertaining of reduction of human casualty cannot be judged from the figures of official entrants only, which at best only tell half the story. So, if there is anything called priority in the delivery of human security system, these poor entrants to the Sundarbans need be provided with a safe sailing boat or dinghy rather than unmonitored and casual experimentation on electrical fencing or battery operated mechanisms. They may be provided with awareness and a sense of security plus proper education so that they may ward off their superstitions and ignorance. Man-killing should be treated as an ecological reaction to the gross over-exploitation of natural resources carried out but he humans for years together and the answer is also to be found by the maintenance of a delicate but complex balance in nature only. There is no other easier alternative.

During the decades 80s and 90s several field Directors have done commendable job. Mention may be made of the names of Pranabesh Sanyal IFS, Vinode Rishi, IFS and P. Sengupta, IFS and others. Sanyal prepared the Management plan of the Tiger Reserve. Under the guidance of all these officers a good number of sweet water ponds were dug and the use of masks/fibre glass shields were set in use by the permit holders. They also experimented placing of human dummy, fitted with electrical devices to scare away the man-eaters from the human beings, obtaining only partial success in all these trials.

Sanyal carried out intensive study of the food habits of the man-eaters. Another achievement was bringing the permit holders under cover of Insurance schemes for undertaking hazardous profession and arrangement for payment of compensation money to the next of kins of death victims and to the injured persons.

More on tiger as viewed by Sy Montgomery

(Source: Spell of Tiger, 1991)

"Wherever it is found, the tiger evokes a reverence, dread and wonder accorded to no other animal."

"Everywhere it roams, the tiger credited with powers beyond those expected of any wordly animal. Nowhere are those powers more fearful than in Sundarbans, largest tidal delta in the world. Here unlike any other on earth, tiger regularly hunts people, hundreds die every year in the tiger's jaws."

"Here the tiger is feared but not hated, here it is worshipped, but not loverd, for here tiger is a sacred creature who ruled an enchanted land."

"In this book the tiger will carry the readers on a spiritual journey to a land where nature and God are one. But we may soon be unable to make, for soon there may be no more tigers to carry is there."

"The tiger must accomplish it drama. It must hunt, it must eat, it must raise its young. And if the Tiger is the guardian of Sundarbans forests, it is because it kills and eats people."

"The tiger's power permeates Sundarbans just as salt flavours its rivers. Most of the tiger does not choose to be seen, just as you cannot see the wind. But you can see what the wind and the tiger have touched."

Sy Montgomery did not fail to mention about the village Amarpur near Gosaba. This village has been designated as 'Vidhaba Palli' or 'widow village' as in each of its 125 families has a woman whose husband or brother or son was killed by a tiger. She mentioned that about the end of 19th century 4218 people were killed by tigers in Sundarbans.

Human fatalities in Sundarbans:

1018 nos. in 28 years i.e. 32.8 persons/year

Honey Collector	–	213 nos.
Fisherman	–	531 nos.
Timber Collector	–	45 nos.
Others	–	229 nos.
Total number of tigers in 1999	–	284 nos.

A UNIQUE PERCEPTION ABOUT MAN-EATING TIGER

No animal on earth has been integrated as a part and parcel of human-life style and influenced their Super-animal conception

Unique Tigers

- Extremely intelligent. Keeps on moving continuously not to leave pug-marks near human habitations.
- Very expert in camouflaging and making no noise while walking.
- Excellent swimmer and very fast runner and expert in acrobatic movements while preying.
- Makes strange sounds while attacking prey animals.

- Diabolical under standing of human behaviour.
- Propensity to kill man mostly of age between (35-45).
- Attack man from behind on the right side of the head, piercing the jugular vein.
- Attack inflicted on man is generally in the morning (7-9 a.m.) and in the evening (3-5 p.m.).
- In general tiger prey during day time.
- Attacks men sleeping in the boat at 11 pm. (tiger swims some distance to reach the boat and jump on it without tilting the boat.
- Eats entire human body including the skull. In 28% cases only corpses could be retrieved.
- They exume dead human body, however deep they are buried.

- Adjustment with drinking of saline water.
- Adjustment with changing habitat conditions, affected by high and low tides.
- Adjustment with habitat full of silt laden soil, infested with crabs, snakes, breathing roots and very dense and low vegetation.

- Tigers break honey-combs and chew honey. It is believed that they smear the body with silt prior to such action as a protective measure against stinging by the bees.

- Tigers survive with damaged livers and kidneys (probably) by drinking of saline water, which might cause physical irritation, enhancing ferocity among man-eaters.

anywhere in the world to such an extent as Sundarbans Tiger. This animal has been unique in shaping social ecology of the people of the region. The inhospitable terrain (mud, tide, salinity, immumerable protruding plant roots, cyclone, storm, crocodile, snakes etc.) has moulded the animal with majestic characteristics, that have branded the animal as most cunning, strong and having supernatural powers, thousand of legends depict the elusive vile character of this magnificent animal that has caused fatality in the lives of poor wood-cutter, honey-collector and fishermen of the terrain, who eke their livelihood in this hostile terrain only. They die in large numbers, yet they come to earn their livelihood without bearing any grudge against the animal, which is unique and a bewildering example of man-animal interaction.

Gunins recite mantras in which they call Banabibi as 'Ma' and seeks her protection, the way she protected them in their mother's wombs. The Gunins address 'Dakakshin Ray' as father, saying that they were entering his domain (the forests) to earn their living and beg to spare them from his wrath.

SOME GUNINS CHANT THE MANTRAS AS FOLLOWS,

"you dwell in the forest,
you, deity of the forest.

4

Sundarbans Estuary World's Unique and a Very Complex Habitat

Sundarban delta has innumerable and complex physiographical parameters. The forests act as nutrient filter and synthesizer of organic matter. The mangrove areas are nursery ground for many species, abode of a wide varieties of vertebrate and invertebrate fauna such a bird, fish, mammals, reptiles, mollusc, Crustacea, bivalve, zoo planktons, phytoplanktons and their interaction.

There are innumerable such islands that make the estuary a complex and unique in character.

Some special features of the areas are:

- Huge swampy area with largest mangrove forest extending towards the east.
- The only place where tiger eats man.
- People get killed by tiger, while they consider it as an object of veneration.
- "The morning mists, the water joining the sky, the tides, the water dissolves the earth – as the self, Hindu mystics say, dissolves into the mind of God. The earth itself, fine, silt clay, came here as rain dissolved fragments of the

The Chart presented below outlines the unique and complex habitat of Sundarban estuary

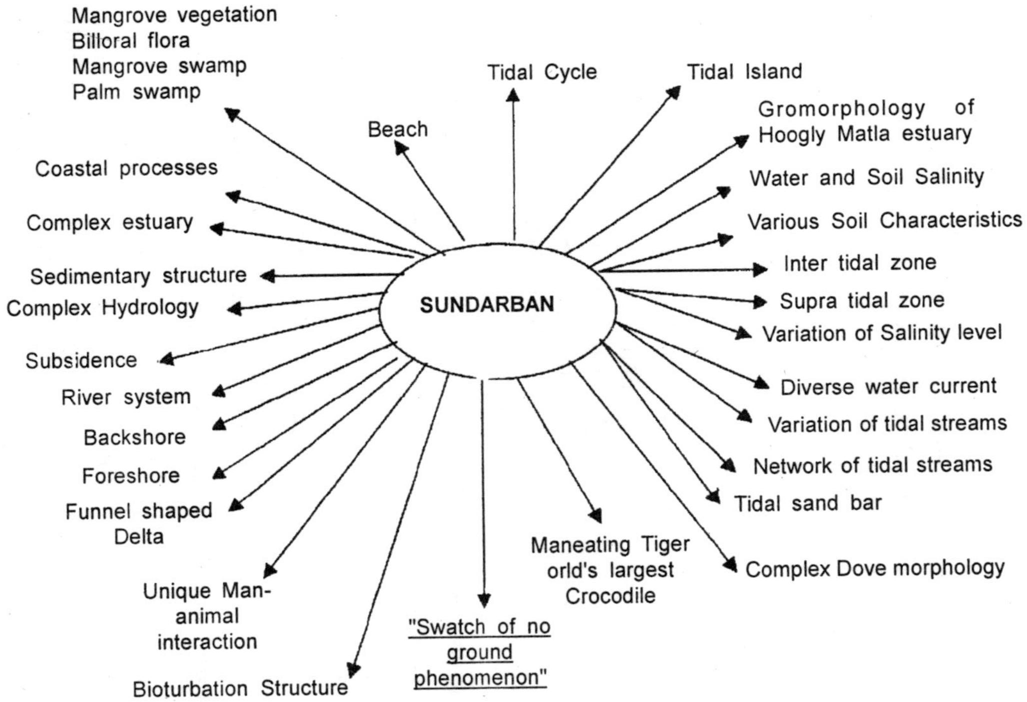

holy Himalayas, carried by the rivers Brahmaputra and the Ganga, themselves considered Gods. The soft mud here looks benign even luscious."

- Here forests and ocean are often indistinguishable, one blending into other like tints in a water colour blue-grey, olive greens and muddy brown.
- "Like the many-headed deities of the Hindu pantheon, it embodies terror as well as peace."
- "Once you put your foot on the mud bank, you know: this is a strange place. Here you know you are at the mercy of something else."
- "Once you leave the wide rivers if you enter the small countries, if you set foot in the forest – and enter a world where the ground sucks you down whole, where the night swallows the stars, and where you know, for the first time, that your body is made of meat."
- Here the cartographers have to redraw their maps at short durations. Nature does not obey rules, fish climbs trees, the animals drink salt water, the roots grow towards the sky instead of down to the earth; tides run in opposite directions in the same creek.

Various broad features of the complex habitats are:

Periodicities	Saline water
Wind effect	Sweet water
Turbidity of water	Thermal dynamics
Estuarine and tidal current	Bioluminescence
Swampy and muddy terrain	Action of chemical physical, mechanical and biological elements, affecting environment.
Abundance of dangerous animals: - Man-eating tiger - Man-eating crocodile - Man-eating fish (shark)	**Forestry operation:** (Collection of fuel-wood, honey, honey comb and Golpata, creation of plantations.
Exclusive beautiful panoramic landscape	Scattered islands, interspersed by natural creeks and nallahs
Inaccessible and in-hospitable terrains.	

SOME SPECIAL CHARACTERISTICS OF THE HABITAT RARELY FOUND ELSEWHERE ARE:

(i) Water

Water in the estuary is saline and turbid. No animal in this terrain has opportunity to drink sweet water. As such, body fluids have to controlled by special osmoregulatory adaptations. Fluids of marine bony fishes are hypnotic; marine fishes drink water and excrete a concentrated urine and excessive salt. The role of saline water in higher terrestrial mammals in the Sundarbans is a virgin field of study. The ferocity in the habit of tiger may be attributed to the damaging liver and kidney developed owing to drinking of saline water for generations.

The characteristic zone of land in between high and low tide (inter-tidal zone) and its extent and depth determines the distribution and nature of animal life in the sub-stratum.

(ii) Periodicities

Many animals of the region (especially the marine) have lunar periodicities in their breeding activities. Marine community does show a variety of rhythmic response to tides. Lunar rhythm does influence salinity, water temperature currents and sediments. Fiddler Crabs (Uca species) have internal biological clock that help them to time their feeding activities. Colour changes on the fiddler crabs are synchronized to dial or tidal cycles.

Preponderance of planktonic flora and fauna has correlation with the salinity, tidal fluctuations, lunar position and the available sunlight. Planktonic fauna, however, has negative correlation with intensity of sunlight and such fauna sink to the bottom (vertical movement).

Study of animals on seashore during daytime only will provide evidence of half the life and behaviour of the communities. All the animals are exposed to three different types of periodism viz., Seasonal, Lunar and Dial (or daily phenomenon).

The activities of tiger relating to human casualties show a definite pattern and further study of facts aided by research will likely reveal some correlation.

Nocturnal collections record significantly more planktons than that of diurnal ones in any water area of the estuary.

(iii) Wind and it's effect on flora and fauna

Strong wind current to which the forests of Sundarbans are exposed during the months from March to September cause immense disturbance to the estuarine plants and animals. Owing to a very low forests most of the animals are exposed to strong wind and rains, water currents and high tides. Small animals are blown over miles: they have to stick fast to the ground or hold tight to tree trunks or in case of smaller animals aggregate to form colonies. A gale of 160 km per hour (not uncommon in Sundarbans) exerts 16 times the force of a wind that is bellowing 40 km per hour. Obviously the formation of muscles and bones have to be tougher than any other habitat to withstand such physical pressure.

The tidal waves sweep over the area twice a day, the tidal current changing its direction after every six hours. The tides penetrate far inland and the maximum rise or fall occurs in March and April. The tidal wave makes its way across the Sundarbans from west to east and consequently the tide changes earlier in the west than it does in the east. In large rivers the velocity of the current usually varies from 3 km an hour near the sea surface to 6 km an hour higher up in the forests. The speed increases further inland on account of gradual constriction of the rivers. The average rise and fall is high in the western rivers than in the east. The average rise and fall near the sea face is about 2.2 m; further up it is 3 m; where the tide has little or no stream to contend against, an average rise and fall of 5-6 m is common.

(iv) Estuary and tidal currents

Estuary is the buffer zone between fresh water of streams and salt water of the sea. Several factors and processes operate in the area. Opposite, unidirectional and oscillating tidal currents meet and exert considerable and complicated force for entire biota. This also affects sedimentation. The tide flow, however, is not constant. It accelerates upto half the height of tide and retards from there downwards.

(v) Salinity

The mixing of salt and fresh water produce a peculiar chemical environment. Diurnal changes of salinity in water creates condition to make physiological adjustments of the animal and plant communities. Salinity also varies widely in vertical and horizontal planes.

Less heat is necessary to get saline water heated and the heat is derived from solar radiation. The temperature of estuary is, therefore, a function of the temperature of entering streams and sea together with tidal stages.

Salinity is a critical factor in the distribution and maintenance of many organisms in the estuary. With the changes of salinity per centages organisms have to adapt osmoregulatory adaptations.

(vi) Thermal Dynamics

Thermal dynamics in estuarine water is complex owing to change in salinity per centage. The form of basin and channel guides the flow of horizontal and vertical distribution of saline water which again is controlled by the tides, currents of water and evaporation.

The present tiger terrain is a negative estuary where the influx of fresh water is not sufficient to maintain mixing. In such estuary oxygen content is low in deep water and the bottom sediments contain considerable organic material.

(vii) Terrain

The Sundarbans is a land of diverse eco-system. This is more than a mere transition zone with varied ecological characteristics and lies between the sea and the inland. The swamps lie close to the sea and have thick forests that are biologically most diverse. Estuarine and marine organism in the area exhibit incredible array of adaptations. Along the shore line thrive thousands of adapted species. The terrain is one of the most naturally fertile in the world. It is also a curiously anomalous tract.

(viii) Bio-luminescence

During the winter months quite often the river water near sea face is found to glow at night. This is due to the pressure of millions and millions of micro-organisms (Dinoflagilates,. Ceratium species, Noctiluca species and Peridium species). These micro organisms contain chlorophyll, carotene and xanthophylls. Entire atmosphere turns into a "Fairy land". Even the forests flanking the streams are frequently illuminated at night by swarms of fireflies.

Do the man-eaters derive extra energy or impetus to prowl at night and search for human prey to pick them up due to this incredible bioluminescence?

(ix) Accessibility

Entire area is practically inaccessible owing to (i) high tide when forests are inundated; (ii) low tide when the shallow rivers do not allow watercrafts to get access; (iii) sharp pointed pneumatophores prevent easy trekking; (iv) soft mud does not allow easy trekking; (v) thick strong stems of regenerations which are impenetrable; (vi) impenetrable masses of stilt root of Rhizophora flanking stream banks and (vii) the rivers being infested by shark, crocodiles etc. and land by snake and man-eaters.

All these make the entire terrain either on land or water impenetrable, difficult to traverse both for animals and human beings.

It becomes immensely difficult for the predators and prey both to easily move in the forests. Similarly the terrain is difficult for human being to make any type of scientific investigation.

The vast expanse of swamps, studded closely with a network of tiny island of mud flats supports a low wooded forest of high density or virgin beauty. The rivers around tiny mud flats, form fantastic sites for eco-biological studies.

Uniqueness is further represented by a few parameters

Why a unique Habitat?

(A broad picture of varied fauna and physiographic features)

Poor fringe population	Poisonous sea-snakes
Man-eating Crocodile	Active delta, where physiography
Man-eating tiger	is fast changing
Man-eating Shark	Peculiar
	- Tidal action
Dolphins/Porpoises	- Frequent storms
in rivers	- Food habits of animals
Preponderence of Oysters,	- Behavioural and food habit
Crabs and other colourful	patterns of monkey, wild boars,
varieties and their abundance	deer and tiger
Wild varieties of fish Species	It is a meso to macro tidal
	estuary
Abundance of Honey Combs	Wide varieties of aquatic birds.
	Typical mangrove forests
	Rare varieties of Turtles.

5

Physiographical Parameters of Sundarbans Delta

These parameters have been broadly outlined in order to impress upon the readers and tourists how wide and complex are each of the facet of the parameters. Each facet has deep scientific reasons of various actions and interactions behind this. For detailed information they may have to refer to various scientific text books.

PHYSIOGRAPHICAL PARAMETERS

The naturalists have to get themselves acquainted with many complex physiographical parameters which have been shown in the charts under Introduction and some more as below:

- Tidal cycle
- Volcanic rock
- Geomorphology
- Soil
- Water
- Severe environmental
- Hazards
- River system
- Mangroves
- Coastal Process
- Complex estuary
- Sedimentary structures
- Hydrology
- Subsidence
- Geological region

GEOMORPHOLOGY OF THE DELTA

- Geomorphology of Matla
- Hooghly complex
- Sedimentary structures in the supratidal dune field
- Wind pattern and wind velocity
- Plant communities
- Bioturbation structures on
- Sandy intertidal flats/muddy flats
- Precipitation
- Tidal Islands
- Dune morphology and stages of formation
- Macrobenthic community and bioturbation structures
- Intertidal Zone

This is an estuarine complex dominated by tide. It's funnel shaped delta with numerous tidal bars and a network of tidal channels. The sand bars of the islands lie more of less perpendicular to the shoreline and parallel to the direction of tidal flow. Channels and sand banks extend for 100 km into the last deltaic island of Sagar. The funnel shaped estuary has been formed through the interactions of the opposing rivers and tidal currents.

The tidal range at the mouth is 5.5m which decreases to 1.0m about 200 km inland from Sagar island. The effect of the tides is felt 300 km inland. The flood tide (High tide) is short in duration (2 to 3 hours) while the ebb flows for the remaining 8 to 9 hours of the 12 hour tide cycle.

QUALITY OF SOIL

- Soil types
- Depositional history
- Soil characteristics
- Soil salinity
- Aeration
- Nitrogen
- Potassium
- Cation exchange capacity
- Minerological composition
- Organic matter
- Available micronutrients
- pH
- Phosphorus
- Sulphur

The soil elements travel a distance of 2880 km. It comprises of fine clay, wilt and sand particles. Characteristic oceanography features like the Hooghly tidal bore occurring at the sea front and the "Swatch of no ground" at the mouth of the river systems cause continuous silt sedimentation upon river banks. The process of sand building occurs as the flow tide scours sand from the adjacent sea bed.

The soil is considered to be of recent origin and have a low oxygen content. The overall pH values vary from 7 to 8 and the humus content is high. In general the soils are found to contain 0.02 to 0.09 per cent nitrogen, 0.06 to 0.1 per cent phosphate, 0.1 to 0.5 per cent patash, 0.0 to 6 per cent Calcium oxide and 0.5 to 0.1 per cent Carbon.

WATER QUALITIES

- Tidal pattern and wind velocity
- Tide level
- Chemical characteristics
  - pH and Salinity
  - Chloride
  - Mineral contents
- Limitation of Photosynthesis by suspended particulate matters
- Extinction coefficient and primary productivity
- Size – fractioned primary productivity
- Variation of salinity level
- Variation of inorganic nutrient exchange
- Seasonal variation in dissolved oxygen and total carbon dioxide
- Semi conservative behaviour of calcium and magnesium
- Behaviour of dissolved silica in relation to planktonic production
- Water current

Aquatic ecosystem forms a complex study. There is immense scope for study of micro and macro plants and animals.

MANGROVE FORESTS – FUNCTIONS

- Nutrient filter and synthesizer of organic matter
- Provide food and act as nursery ground for terrestrial and acquatic animals

- Acts as a buffer between land and sea
- Have high energy production capacity towards estuarine and marine fertility where a rich and diverse life form survives
- Stabilize soil
- Many floral and faunal species have wide distribution
- Most of the fauna are detrivorous; only some in fauna which are omnivorous
- Diverse habitat concern trees, soil, water, mud flat, clay flat, intertidal zone etc.
- Zoo-plankers, Zoo benthos, vertebrates and invertebrates act as consumers.
- Nursery ground for many species

SUBSIDENCE

The "Swatch of no Ground" phenomenon is special feature of the area.

This phenomenon is held responsible for frequent subsidence of some parts of the Sundarbans. The area is marked South of Raimangal-Malancha estuary between 21° – 22′ latitude in the Bay of Bengal. Oldham (1893) noted that "in the sea outside the middle of the delta there is a singularly deep sea area known and marked in the charts as 'Swatch of no ground" in which surroundings which are from 5-10 fathoms all round, change almost suddenly to 200 and even to 300 fathoms."

Fergusson (1863) opined that this feature was formed due to interaction of eastward and westward currents in the Sundarbans which do not allow river carried silt to settle in the area. The silt is carried further south to form new islands.

It is believed that all over the estuaries this phenomenon occur causing unsettled condition due to the existence of this sub-marine area.

The coastal landform is compassed of several elements as follows:

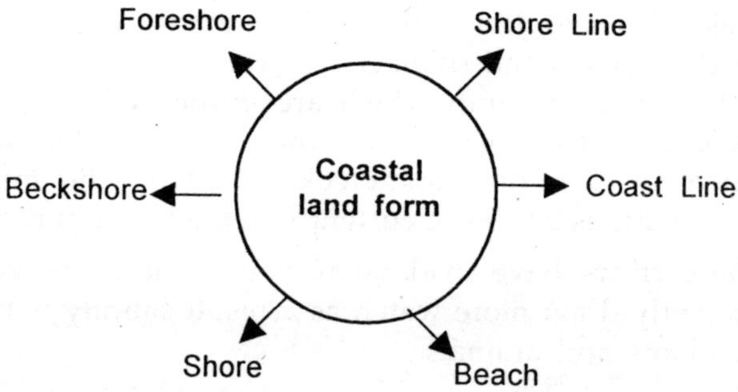

The intertidal zone used to get flushing of the Ganges till 12th century. But during 12th to 16th century the basin tilted eastward into river Padma in Bangladesh. The present Indian Sundarbans do not get enough fresh water. The following rivers interect the delta:

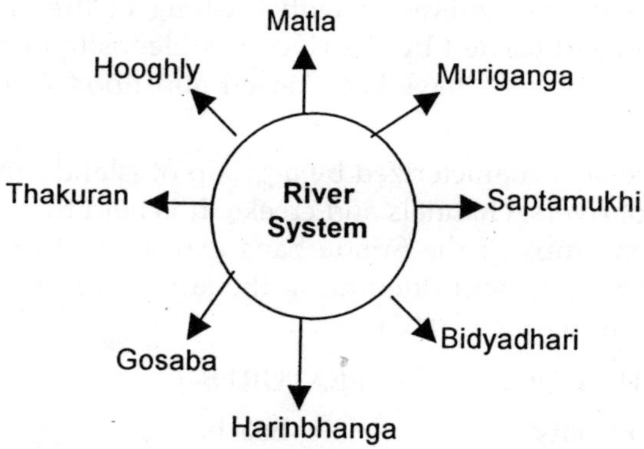

All these together form the Hooghly Matla estuarine complex.

The word 'Estuary' is derived from latin word 'aestuarim' which means total. Some have defined estuary as a semi-enclosed coastal body water which has a free connection with the open sea and within which sea water is measurably diluted with river water. Others define it as an inlet of the sea reaching into a river valley as far as upper limit of the tidal rise. So estuary is a complex mixing zone between forest and sea water, producing a wide range of brackish water of intermediate salinities.

In this aquatic environment a number of chemical interactions occur between suspended water and the dissolved constituents of a succession of fresh and brackish and marine water.

Major Rennel's Map prepared in 1779 shows cris-cross of hundreds of rivers and rivulets in lower Bengal. The rivers run from the north to south which are influenced by tides. The channels connecting these rivers flow east or west. Innumerable silt laden rivers, rivulets and creeks cris- cross the entire delta and have various grades of current flowing through their water.

These rivers have inadequate supply of fresh water and consequently draw more water, as a result salinity is high that affects plants and animals.

All the rivers in Sundarbans are tidal. Average tidal amplitudes in the estuary vary from 3.5 m to 5 metres. Highest amplitudes occur in July-August which is lowest in December-January.

Premature reclamation of flood plan and erection of embankment have interfered with flushing of the plains. As a result sediment carried by the rivers are deposited on the river beds. The river beds have been raised and flood incidence has increased.

The delta is characterized by a group of islands and a dense network of rivers, channels and creeks. It is held that in the past the tidal swamps of the Sundarbans extend up to the base of the Rajmahal hills and fitted up in the latter part of the Tertiary period. (Furgusson – 1803).

SOME PHYSIOGRAPHIC FEATURES

- Tidal flats
- Tidal creeks
- Coastal dunes
- Intertidal mudflat
- Back dunes
- Beaches
- Estuaries
- Mangrove swamps
- Saltwater

SOME COASTAL PROCESSES

- Microtidal cycle
- Long shore current (Post monsoon)
- Tropical cyclone
- Wave parameters
- Long shore current (Pre monsoon)
- Macro-tidal cycles

FEATURES OF SUNDARBANS ESTUARY

(I) Salinity and Tidal Height

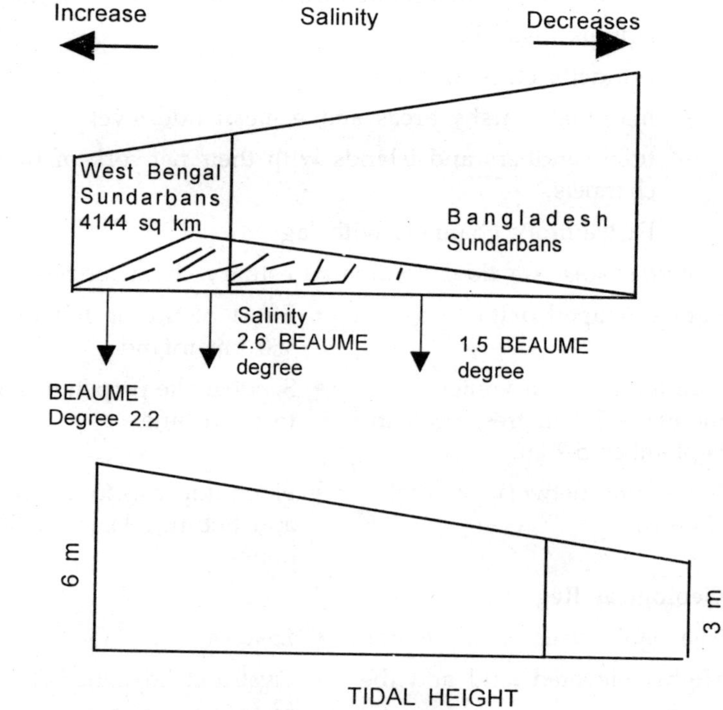

(II) SALINITY AND TREE GROWTH

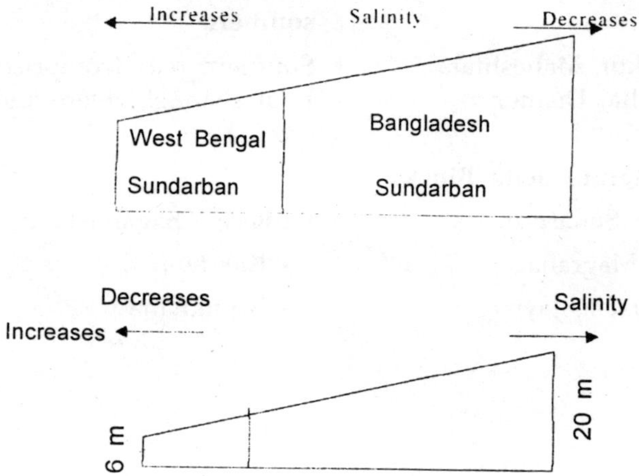

GEOMORPHIC UNITS

Some spectacular geomorphic units include

- sub-aerial and sub-aquous leaves, splays and tidal flats.
- aquous distal bar.
- pro-delta clays and silt.
- marginal marshy areas above mean tide level.
- tidal sandbars and islands with their network of tidal channels.
- Distributory channels with lag.

Sundarbans is a tidal dominated estuary characterized by:

• Funnel shaped delta	• Effect of tide is felt over 300 km inland
• Tide level low in winter months - 3-4 metres, peak in September 5-7 m.	• Sandbars lie perpendicular to shore line
• Numerous network of tidal channels	• Flood tide lasts for 2-3 hrs. and Ebb-tide lasts for 8-9 hours
• **Geological Regions**	
• **The stable Delta on the North**	• **Eastern**
• Higher elevated land and the	• Piyali and Bidyadhari plains
• Land of sluggish rivers and	• Bhangar and Canning
• **Western Hooghly side flat**	• **The Active Delta of the southern**
• Thakurpukur, Maheshtata Budge, Falta, Diamond Harbour, Kulpi	• Southern part (comprises tidal channel, rivers and creeks)
• **Middle Mature delta Blocks**	
• Bishnupur, Sonarpur,	• Western - Sagar island
• Baruipur, Magrahat,	- Kakdwip
• Mandir bazaar, Jaynagar	- Namkhana

- Middle
  - Mathurapur
  - Patharpratima
- Eastern
  - Kultali
  - Basanti
  - Gosaba

The Coastal areas are more productive than the open sea. In this zone sizable populations of phytoplankton-feeding fish, although majority of the fish are carnivores. This zone is very significant for marine fish productivity.

The sources of sediments is the estuary are diverse and are supposed to be derived from the following:

- offshore sediments derived by erosion,
- offshore sediments carried by tidal circulation,
- run-off from land,
- biogenic detritus from estuarine organisms,
- sediments derived by pellet making organisms.

6

Mangroves - What They Are?

The word 'Mangrove' might have been derived from Portugese word 'Mangrove' which refers to mangrove community as a whole. The word refers to community of trees, shrubs and herbs growing in a particular intertidal ecosystem of the tropics.

They form a belt, broad or thin, of small sized trees in mangrove swamps which forms a spectacular environment of the estuary. The humid tropical and sub-tropical climatic factors help development of dominant mangrove swamps. These forests protect the coast. Vegetation zonation and plant succession is best inhibited in such forests. The mangrove roots both subterranean and aerial (pneumatophores) stabilize sediments that help formation and growth of tidal islands.

Some salient characters of mangrove areas are:

- Soil is black coloured and sticky
- Most fertile soil for microbial activities; Cynophyceae (blue-green algae) flourish well in loose saline sedimented soil humus.
- Chlorophyceae (green algae) prefers soil, rich in nitrates and phosphates,
- Soil humus supplies nitrates required for Chlorophyceae.

- Animal excreta enriches nitrate and phosphate of the soil.

There is abundance of Hydrogen sulphide in humus soil:

- Bacteria, green-algae, blue-green algae and diatoms act as nitrogen fixing and sulpher reducing agent in tidal soil.
- High per centage of Calcium ion in mangrove soil reduce/prevent adverse effect of sodium ion.
- Microbes decompose leaves, roots and in the process organic carbon increases.

MANGROVE PLANTS

Naskar KR. has done exclusive work on the mangroves. He considers them to be distributed in three categories which are as follows:

Major elements (Genera) :

Rhizophora, Bruguiera, Cereops Kandelia, Avicennia, Sonneretia, Lumnitzera and Nipa.

Minor elements (Genera):

Xylocarpus, Aglaia, Aegiceros, Ex-coecria, Aegialitis, Heritiera, Atlantia, Brownlowia, Acanthus, Acrostichium and Scyphiphora.

Mangrove Associated (Genera):

Hibiscus, Thespesia, Cynometra, Dalbergia, Derris, Phoenix and others (28 spp.).

FOREST TYPES

Champion and Seth (1968) have classified the forests into 5 (five) following types

- Mangrove scrub
- Mangrove forest
- Salt water Mixed Forest
- Brackish water Mixed Forest
- Palm Swamp

Principal species are:

Ceriops tagal, C. decandra, Excoecaria agallocha, Avicennia sp.

Various associate formations are:

- Heritiera type – Heritiera and Phoenix
- Low mangrove type : Heritiera and Excoecaria, Heritiera – Excoecria – Ceriops
- Pure Phoenix – (Common)
- Pure Ceriops – (Common)
- Avicenia – Porteresia
- Avicennia – Sonneretia
- Pure Excoecaria (Common)

Blocks and Compartments

For better conception about the area a map and some charts/ lists are presented for an overall guidance of the ecoeducation tourists.

Under 24-Parganas (South) Division

Herobhanga (8)

Thakuran (4)

Chulkati (8)

Dhupibhasani (8)

Ajmalmari (12)

Lothian

Saptamukhi (2)

Under Tiger Project

Jhilla (6), Arbesi (5), Khatuajhuri (3), Chandkhali (4), Chamta (8), Matla (4), Gona (4), Bagmara (8), Harinbhanga (3), Gosaba (4) Chota hardi (4), Mayadwip (5), Netidhopani (3), Prikhali (7), Panchamukhani (5)

The above list presents the names of various blocks of Sundarbans forests number of compartments being enclosed in brackets.

FLORAL RESOURCES

Sundarbans mangrove does not have many plant species. In the halophytic habitat there are hardly forty species (but there cannot be any fixed number of plants to pin point). Several associates of tree mangrove are also provided in the list of plants.

In the reclaimed area many species have been recorded (Naskar and Guha Bakshi 1987). This author has given a lists of plant in his UNESCO publication which include Naskar-Guha Bakshi finding. For the present purpose a list of plants presented by P. Sanyal (Calcutta Env. Society 1997) is reproduced in a modified form which gives a check list with update nomenclature as follows:

Families & Genera of Mangroves & the Mangrove	Species Names of the Mangroves/ Mangrove	Local name Vernacular name
RHIZOPHORACEAE		
Rhizophora	*R. apiculata*	Bhora, Garjan
(All are Trees)	*R. mucronata*	Khamu, Garjan
Bruguiera	*B. gymnorrhiza*	Kakra
(Most are Trees/ Few	*B. sexangula*	Kakra
are Shrubs)	*B. cylindrica*	Bakul Kakra
	B. parviflora	Bakul Kakra
Ceriops	*C. decandra*	Jhamti Goran
(Small Trees/Shrubs	*C. tagal*	Mat Goran
Kandelia	*K. candel*	Goria
(Small Tree)		
AVICENNIACEAE		
Avicennia	*A. alba*	Kal-ban/bain/bani
(Small-Medium Tree)	*A. officinalis*	Jat-ban/bain/bani
	A. marina	Peyara-ban/ bain/ bani
SONNERATIACEAE		
Sonneratia	*S. alba*	Sada Chak-Keora
(Medium-Tall Tree)	*S. caseolaris*	Chak-Keora
	S. griffithii	Ora
	S. apetala	Keora
COMBRETACEAE		
Lumnitzera	*I. racemosa willd*	Kripa/Kripal
(Shrub/Small Tree)		

(contd.)

Families & Genera of Mangroves & the Mangrove	Species Names of the Mangroves/ Mangrove	Local name Vernacular name
ARECACEAE		
Nypa Steck	*N. fruticans (Thunb.)*	Golpata
MALVACEAE		
•*Hibiscus*	*H. tiliaceus*	Ban-Kapas
(Medium Tree)	*H. tortuosus*	Ban-Kapas
•*Thespesia*	*T. populnea*	Paras
(Medium Tree)	*T. populneoides*	Paras
CAESALPINIACEAE		
•*Caesalpinia*	*C. bonduc*	Nata-Karanja
(Twining Shrub)	*C crista*	Shingri-Lata
•*Aizoaceae*		
Sesuvium	*S. portulacastrum*	
FABACEAE		
•*Dalbergia*	*D. spinosa*	Chulia Kanta
(Small Tree)		
•*Derris*	*D. scanden*	Noa-Lata
	D. trifoliata	Pan-Lata
	D. indica	Karanja
ARERACEAE		
Phoenix	*P. paludosa*	Hental
COMBRETACEAE		
Terminalia	*T. catappa*	Kat Badam
(Tree)		
EBENACEAE		
Diospyros	*D. ferrea*	Ban Gab
(Tree or shrub)		
MELIACEAE		
Xylocarpus	*X. granatum*	Dhundal
(Medium-Tall Tree)	*X. mekongensis*	Pasur
Aglaia	*A. cucullata*	Amur

Families & Genera of Mangroves & the Mangrove	Species Names of the Mangroves/ Mangrove	Local name Vernacular name
MYRSINACEAE		
Aegiceras	*A. corniculatum*	Khalsi
(Shrub-Small Tree)		
•*Euphorbiaceae*		
Excoecaria	*E. agallocha*	Gneo/Gneoa
(Medium Tree)	*E. indica*	Gneo/Gneoa
AEGIALITIDACEAE		
(nom.alt.Plumbaginaceae)		
Aegialitis	*A. rotundifolia*	Tora
STERCULIACEAE		
Heritiera	*H. fomes*	Sundari
(Tall Tree)		
RUBIACEAE		
Scyphiphora	*S. hydrophyliacea*	Tagri Bani
(Shrub)		
PTERIDACEAE		
•*Acrostichum*	*A. aureum*	Hudo fern
ACANTHACEAE		
•*Acanthus*	*A. ilicifolius*	Harkoch
(Shrub/Twiner)	*A.* volubilis	Lata-Hargoja
RUTACEAE		
Atalantia	*A. correa*	Ban Lebu
(Shrub)		
TILIACEAE		
Brownlowia	*B. tersa*	Lata-Sundari
(Shrub)	*(=B. lanceolata)*	
SAPINDACEAE		
Allophyllus	*A. cobbe*	

(*contd.*)

Families & Genera of Mangroves & the Mangrove	Species Names of the Mangroves/ Mangrove	Local name Vernacular name
ASCLEPIADACEAE		
•*Sarcolobus*	*S. carinatus*	Baoli Lata
(Twiner/Shrub	*S. globosus*	Baoli Lata
•*Pentatropis*	*P. capensis*	-
CHENOPODIACEAE		
•*Suaeda*	*S. maritima*	Girc-sak
(Herbaceous)	*S. nudiflora*	Girc-Sak
Salicornia	*S. brachiata*	-
•(*Herbaceous*)		
AMARYLLIDACEAE		
Crinum	*C. defixum*	Sukh-Darshan
(Herbaceous)		
ARACEAE		
•*Cryptocoryne*	*C. ciliata*	Kerali
VERBENACEAE		
•*Clerodendrum*	*C. inerme*	Ban Jui
(Shrub)		
Premna corymbosa		
RUPPIACEAE		
•*Ruppia*	*R. maritima*	Nona Jhanji
TAMARICACEAE		
•*Tamarix*	*T. dioica*	Nona Jhau
(Shrub-Small Tree)	*T. galica*	Nona Jhau
POACEAE		
•*Porterasia*	*P. coarctata*	Dhani Ghas
•*Myriostachya*	*M. wightiana*	Nalai
•*Aeluropus*	*A. lagopoides*	Nona Durba

Families & Genera of Mangroves & the Mangrove	Species Names of the Mangroves/ Mangrove	Local name Vernacular name
SOLANACEAE		
Solanum	*S. trilobatum*	Nona Begun/Kat
CYPERACEAE		
Cyperus	*C. jaranicus*	Begun
LAURACEAE		
Cassytha	*C. filiformis*	Akash Bel
LORANTHACEAE		
•*Macrosolen*	*M. cochinchinensis*	Chotomanda
•*Viscum*	*V. moncicum*	Manda
	V. orientale	Manda
RUBIACEAE		
Hydrophyllax	*H. maritima*	
CONVOLVULACEAE		
Ipomoea	*I. pes-caprae*	Chagal Kuri

[NB. Some Botanists put the species marked • as mangrove associates].

LIST OF ALGAE OF SUNDARBANS

This list compiled by the Calcutta Wildlife Society lovers only prophetic and Benthic algae. This list must be supplemented by phytoplanktonic algae and those recorded by CIFRI from brakish water over a long period. The author has recorded the planktonic algae in appropriate chapters having quoted in CICFRI's publications.

Name of Division, Class, Order and Family

CHLOROPHYTA

1. *Enteromorpha clathrata* Periphytic
2. *Enteromorpha compressa* Periphytic
3. *Enteromorpha intestinalis* Periphytic

(*contd.*)

4.	*Enteromorpha prolifera*	Periphytic
5.	*Enteromorpha lubulosa*	Periphytic
6.	*Ulva lactuca*	Periphytic
7.	*Ulva patengensis*	Periphytic
8.	*Ulva* sp.	Benthos
9.	*Chaetomorpha acrea*	Periphytic
10.	*Chaetomorpha bischyona*	Periphytic
11.	*Chaetomorpha gracilis*	Periphytic
12.	*Chaetomorpha linum*	Periphytic
13.	*Lola capillaries*	Periphytic
14.	*Lola implexa*	Periphytic
15.	*Lola tortusa*	Periphytic
16.	*Rhizoclonium grande*	Periphytic
17.	*Rhizocionium hookeri*	Periphytic
18.	*Rhizoclonium kerneri*	Periphytic
19.	*Rhizoclonium riparium*	Periphytic
20.	*Cladophora echinus*	Periphytic
21.	*Cladophora patentiramea*	Periphytic
22.	*Cladophorella sundarbanensis*	Periphytic
23.	*Cladophoropsis robusta*	Periphytic
24.	*Spirogyra* sp.	
25.	*Dictyosphaeria cavernosa*	Periphytic
26.	*Boodlea composita*	Periphytic
27.	*Bryopsis indica*	Periphytic
28.	*Caulerpa caetoides*	Benthos
29.	*Caulerpa racemosa* var. *clavifera*	Benthos
30.	*Caulerpa racemosa* var. *occidentalis*	Benthos
31.	*Caulerpa racemosa* var. *turbinata*	Benthos
32.	*Caulerpa racemosa* var. *uvifera*	Benthos
33.	*Caulerpa scalpelliformis*	Benthos
34.	*Caulerpa sertularioides*	Benthos

35.	*Halimeda discoidea*	Benthos
36.	*Characium* sp.	Epiphytic
37.	*Coleochaete* sp.	Epiphytic
38.	*Ulothrix* sp.	Periphytic
39.	*Acetabularia calyculus*	Periphytic
CHRYSOPHYTA		
40.	*Vaucheria erythrospora*	Periphytic
41.	*Vaucheria longicaulis*	Periphytic
42.	*Vaucheria mayyamadensis*	Periphytic
43.	*Vaucheria prescottii*	Periphytic
44.	*Vaucheria pronosperma*	Periphytic
PHAEOPHYTA		
45.	*Ectocarpus erythrospora*	Periphytic
46.	*Ectocarpus rhodochortonoides*	Periphytic
47.	*Ectocarpus sp.*	Periphytic
48.	*Giffordia conifera*	Periphytic
49.	*Giffordia duchassaigniana*	Periphytic
50.	*Giffordia irregularis*	Periphytic
51.	*Giffordia mitchella*	Periphytic
52.	*Giffordia rallisiae*	Periphytic
53.	*Giffordia thyrsoideus*	Periphytic
54.	*Feldmannia columellaris*	Periphytic
55.	*Feldmannia elachistaeformis*	Periphytic
56.	*Feldmannia indica*	Periphytic
57.	*Feldmannia vaughani*	Periphytic
58.	*Sphacelaria furcigera*	Benthos
59.	*Sphacelaria tribuliides*	Benthos
60.	*Dictyota bartayresii*	Benthos
61.	*Dictyota dichotoma*	Benthos
62.	*Dictyota divaricata*	Benthos
63.	*Dictyota friabilis*	Benthos

(contd.)

64. *Dictyota patens*	Benthos
65. *Dictyopteris australis*	Benthos
66. *Lobophora variegata*	Benthos
67. *Padina australis*	Benthos
68. *Padina tenius*	Benthos
69. *Padina gymnospora*	Benthos
70. *Padina pavonica*	Benthos
71. *Padina sanctae-crucis*	Benthos
72. *Padina tetrastromatica*	Benthos
73. *Padina vickersi*	Benthos
74. *Myriactula arabica*	Benthos
75. *Chroospora implexa*	Benthos
76. *Colpomenia sinusa*	Benthos
77. *Hydroclathrus clathrus*	Benthos
78. *Rosenvingea floridane*	Benthos
79. *Rosenvingea intricata*	Benthos
80. *Rosenvingea orientales*	Benthos
81. *Rosenvingea sanctae-crucis*	Benthos
82. *Sargassum carpophyllum*	Benthos
83. *Sargassum heterosystum*	Benthos
84. *Sargassum ilicifolium*	Benthos
85. *Sargassum piluliferum*	Benthos
86. *Sargassum vulgare*	Benthos
87. *Sargassum tenerrimum*	Benthos
88. *Sargassum wightii*	Benthos
RHODOPHYTA	
89. *Goniotrichum alsidi*	Benthos
90. *Erythrocladia subintegra*	Benthos
91. *Erythrocladia carnea*	Benthos
92. *Acrochaetium crassipes*	Benthos
93. *Liagora ceranoides*	Benthos

94.	*Actinotrichia fragilis*	Benthos
95.	*Galaxaura fastigiata*	Benthos
96.	*Scinaia complanata*	Benthos
97.	*Gelidiella tenuissima*	Benthos
98.	*Gelidium pusillum*	Benthos
99.	*Jania adhaerens*	Benthos
100.	*Jania ungulata*	Benthos
101.	*Amphiroa fragilissima*	Benthos
102.	*Meloresia conferuicola*	Benthos
103.	*Hypnea boergesenii*	Benthos
104.	*Hypnea coenomyce*	Benthos
105.	*Hypnea musciformis*	Benthos
106.	*Hypnea pannosa*	Benthos
107.	*Hypnea valentiae*	Benthos
108.	*Sarconema furcellatum*	Benthos
109.	*Catenella impudica*	Benthos
110.	*Catenella repens*	Benthos
111.	*Polysiphonia denudata*	Periphytic
112.	*Polysiphonia mollis*	Periphytic
113.	*Pterosiphonia pennata*	Periphytic
114.	*Compsopogon coeruleus*	Periphytic
115.	*Acanthophora specifera*	Periphytic
CYANOPHYTA		
116.	*Xenococcus chaetomorphae*	Periphytic
117.	*Xenococcus cladophorae*	Periphytic
118.	*Aphanotheca castagnii*	Periphytic
119.	*Merismopedia glauca*	Periphytic
120.	*Gloeocapsa calcarea*	Periphytic
121.	*Gloeocapsa montana*	Periphytic
122.	*Gloeocapsa kuetzingiana*	Periphytic
123.	*Gloeocapsa rupestris*	Periphytic

(*contd.*)

124.	*Gloetheca samoensis*	Periphytic
125.	*Gloetheca membranacea*	Periphytic
126.	*Merismopedia tennissium*	Periphytic
127.	*Aphanocapsa thermalis*	Periphytic
128.	*Aphanocapsa roeseana*	Periphytic
129.	*Aphanocapsa pulchra*	Periphytic
130.	*Microcystis litoralis*	Periphytic
131.	*Microcystis bengalensis*	Periphytic
132.	*Microcystis pulverea*	Periphytic
133.	*Myxosarcina spectabilis*	Periphytic
134.	*Johannesbaptistia pellucida*	Periphytic
135.	*Chamaesiphon curvatus*	Periphytic
136.	*Dermocarpa hemisphaerica*	Periphytic
137.	*Stichosiphon sansibaricus*	Periphytic
138.	*Raphidiopsis indica*	Periphytic
139.	*Oscillatoria amoena*	Periphytic
140.	*Oscillatoria bonnemaisonii*	Periphytic
141.	*Oscillatoria margaretifera*	Periphytic
142.	*Oscillatoria martini*	Periphytic
143.	*Oscillatoria subbrevis*	Periphytic
144.	*Oscillatoria tenuis*	Periphytic
145.	*Lyngbya allorgei*	Periphytic
146.	*Lyngbya confervoides*	Periphytic
147.	*Lyngbya hieronymusii*	Periphytic
148.	*Lyngbya lutea*	Periphytic
149.	*Lyngbya magnifica*	Periphytic
150.	*Lyngbya semiplana*	Periphytic
151.	*Hydrocoleum cantharidosum*	Periphytic
152.	*Schizothrix lamyi*	Periphytic
153.	*Microcoleus chthonoplastes*	Periphytic
154.	*Plectonema wollei*	Periphytic

155.	*Anabaena variabilis*	Periphytic
156.	*Crinalium magnum*	Periphytic
157.	*Scytonema siculum*	Periphytic
158.	*Calothrix conferviflora*	Periphytic
159.	*Calothrix parasitica*	Periphytic
160.	*Calothrix cotarenii*	Periphytic

7

Vertebrate Fauna of Sundarbans

The Sundarbans mangroves are rich in terrestrial, aquatic, and avian faunal species. The forests and waterways provide dwelling places, habitats, breeding areas and refuge for a wide variety of species including 8 species of amphibians, 57 species of reptiles, 161 bird species and 40 species of mammals, many of which are endangered in other parts of the world.

A check list and a small account of the fauna is given below:

(A) MAMMALS

Suncus murinus House Shrew (or Grey Musk Shrew)

Manis pentadactyla (Pangolin)

Macaca mullata (Rhesus Monkey)

Funambulus pennanti (Five-stripped Squirrel)

Mus booduga (Little Indian Field Mouse)

Rattus rattus (House Rat)

Bandicota bengalensis (Lesser Bandicoot Rat)

Bandicota indica (Large Bandicoot Rat)

Hystrix indica (Indian Crested Porcupine)

Platanista gangetica (Gangetic Dolphin)

Neophocaena phocaenoides (Little Porpoise or Black Finless Porpoise)

Orcella brevirostris (Irrawady Dolphin)

Sotalia plumbea (Plumbeous Dolphin)

Canis awreus (Jackal)

Vulpes bengalensis (Indian Fox)

Felis chaus (Jungle Cat)

Felis bengalensis (Leopard Cat)

Panthera tigris (Tiger)

Paradoxurus hermaphoroditus (Toddy Cat)

Viverricula indica (Small Indian Cevet)

Herpestes edwardsi (Common Grey Mongoose)

Herpestes auropunctatus (Small Indian Mongoose)

Herpestes palustris (Marsh Mongoose)

Lutra lutra (Common Otter)

Sus scrofa (Wild Boar)

Cervus axis (Spotted Deer)

"Terrestrial brackish water and marine fauna constitute the many faunal components in this ecosystem. Mangrove fauna occupy, in general three different biomes, namely, littoral or super-littoral forests, intertidal mudflat and estuary. The littoral or supralittoral forest biome is typically a terrestrial environment which includes both aerial and arboreal forms and the soil inhabitants. Arboreal animals and soil dewellers may be categorized as epifauna and infauna. The intertidal mudflats are essentially semiterrestrial or semiaquatic habitat supporting mainly the soil forms and the benthos. While the estuary is inhabited by aquatic forms comprising of plankton and nekton.

The pattern of distribution of animals in mangrove ecosystem is influenced by the substratum, salinity tidal amplitude, vegetation, light, temperature etc.

The supra-littoral forest habitat includes areas where water may or may not reach at all and is essentially densely covered with halophytes. It offers forest floor, roots stems, branches and leaves of trees as the abode. The intertidal mudflat proper is recognized as areas having periodical inundation and exposure i.e. either exposed or submerged with tidal flush for certain period of the day and is sparsely vegetated or with no vegetation at all in the lower littoral zone. The third biotope is the estuary or the creek which has characteristically low-saline brackish water system."

[**Source**: Modified from Mandal and Nandi of ZSI]

(B) MAMMALS (BATS)

Chiroptera (Bats of Sundarbans

Cynopterus sphinx (short-nosed fruit bat)	*Megadrma lyra* (Indian vampire bat)
Pteropus giganteus (Indian flying fox)	*Rhinopoma hardwickii* (Lesser mouse-tailed bat)
Rousettus leschensulti (leschenault's rousette fulvous fruit bat)	*Coeciops frithi* (Tailless leaf-nosed bat)
Hippasideros lankdiva (Cave bat)	*Hespirotenus tickelli* (Indian pigmy pipistrella)
Pipistrella coromandra (Serotine bat)	*Scotophilus kukli* (Common yellow bat)

Arboreal invertebrate Fauna

Butterflies	Fire flies (Pleroptyx sp.)
Moths	Ants
Dragon flies	Mosquitoes (Aedes, Anopheles
Bees	and Culex spp.)
Beetles	Wasps
Slug (Onchidium)	

Other invertebrate fauna

Bivalve	**Mollusc**
Crassostrea sp.	Cussiduata sp.

Teredo sp.	Cerethidua obtuse
Balanas sp.	Ellobium arisjudae
Bankia sp.	E. gangeticum
Nausitora sp.	Littorina melanostoma
Bactoronophore sp.	Nerita violacea
	Orchidium undulata

LIZARDS AND REPTILES

In water and on land Sundarbans have the unique distinction of an assemblage of a wide number of species of lizards, snakes and turtles and tortoises. The researchers will have an array of subject to study of these fauna, their interactions with other animals and man.

For the convenience of the tourists lists of some conspicuous species of the above animals have been presented below:

Of the 57 species of reptiles found in the Sundarbans, 12 belong to the order Chilonia, 44 belong to Squamata and 1 to Crocodilia.

The order Squamata has the largest of representatives, including Gekko geko, the Indian chamaeleon (*Chamaeleon zeylanicus*), the water monitor Varanus salvator, the yellow monitor (Varanus falvescens) and terrestrial aquatic snake spekcies. The water monitors feed on crustaceans, mollusks, bird's eggs and fledgings.

Tree snakes of the Sundarbans include the Indian bronze back (*Dendrelaphis tristis*) and the Wolf snake (*Lycodon anlicus*). Other snake found are the Indian python, Rock python Russell's sand boa, Rat snake, Keelbacks (*Acrodhordus granulatus*) krait cobra and vipers. A number of aquatic snakes as the Tinket snake and sea snakes such as the Dog-faced water snake (*Cerberus rhynchops*) are common.

Fresh water tortoises of terrapins and marine turtles belong to the order Chilonia. Turtles and terrapins are represented by the endangered river terrapin (*Batagur baska*), Olive Ridley Turtle (*Lepidochelys olivacea*), Ganges softshelled turtle (*Trionyx gangeticus),* peacock marked soft-shelled turtle (*Trionyx hurum*), Gray and Indian Flapshelled turtle (*Lissemys punctata*).

The commercial value of the shells and other parts of many reptilian species has had a serious impact on their continued survival in the Sundarbans. A number of conservation and protection efforts have been made, including the establishment of a crocodile farm and hatcheries for the Olive Ridley turtle.

Reptiles : Turtle and Tortoises

Turtles are compactly built reptiles with bony shells, which is the part of the skeleton protecting both the back and underside. They live in all parts of the earth and the oceans. Aquatic turtles do not feed exclusively on other animals. In many species the proportion of vegetable and animal food are approximately equal. The genera Batagur, Kachuga, Hardella feed primarily on plants so also the green turtle Celonia mydas. The turtles have a definite sense of time, their daily schedule is well regulated. The life of the turtle follows a daily as well as annual rhythm.

Turtle and Tortoises

Chelonia mydas (Green furtle)	*Batagur baska* (River terrapin)
Lepidochelys olivacea (Olive Ridley turtle)	*Trionyx gangeticus* (Gangetic soft shelled turtle)
Lissemys punctata (Indian flapshelled turtle)	*T. hurum* (Peacock marked soft shell turtle). Also occurs in the marginal areas.
Kachnga kachuga (Reed crowned roofed turtle)	*Kachuga tecta* (Roofed turtle)
Morenia ocetalla (Yellow turtle)	*Melanochelys tricarinata* (Three keeled and tortoise)
Chitra indica (Narrow headed softshell turtle)	*Geoclemys hamiltonii* (Spotted pond turtle)

The naturalists will find a wide scope to study on biology of a few species of turtle and estuarine crocodile in Forest Deptt. Research center at Bhagabatour near Namkhana.

Reptiles: Snakes ot Sundarbans

Typhlops porrectus

Typhlops braminus

Python molurus, (Indian Python)

Eryx conicus, (Russell's Sand Boa)

Acrochordus granulatus, (Thinket Snake)

Elaphe helena

Ptyas mucosus, (Rat Snake)

Oligodon arnensis

Ahaetulla nasustas, (Green whip Snake)

Ahaetulla myctarijans

Lycodon aulicus, (Common wolf Snake)

Natrix stolata, (Checkered Keelback)

Atretium schistosum, (Olivaceous Keelback)

Enhydris enhydris

Cerberus rhynchops

Gerardia frevostiana

Dendrelaphis acactullah

Dendrelaphis tristis

Bungarus careuleus, (Common Indian Krait)

Bungarus fasciatus, (Banded Krait)

Naja naja, (Common Cobra)

Ophiophagus hannah, (King Cobra)

Praescutata viperina

Enhydrina schistose

Hydrophis nigrocinctus

Hydrophis obscurus

Hydrophis caerulescens

Microcephalophis gracilis

Microcephalophis contoris

Vipera russelli, (Russell's viper)

Trimeresurus erythrurus, (Pit Viper)

The Estuarine crocodile (*Crocodylus porosus*)

The salt water of Estuarine crocodile (*Crocodilus porosus*) has been sighted throughout the 2500 sq. km of the Sundarbans although the species appears to be concentrated mostly in the Saptamukhi, Thakuran, Choto Naubanki and Mechua Khaal areas. In recent years, sightings of crocodiles have become less frequent, perhaps indicating a decline in number due to indiscriminate killing. Over the last couple of decades, an average of six people have been killed annually by crocodiles, although, in recent years, reported incidents have been fewer.

It has been noted that a number of migratory birds such as the common sandpiper, wood sandpiper, green sandpiper, swamp duck and the Turnstone have not left the Sundarbans area after the winter season (Mandal and Nandi, 1989). This change in behaviour should be observed to determine any long term effect on the species.

The populations of marine turtles and reptiles are also threatened, although hatcheries for the Olive Ridley turtle and the Crocodile farm have helped alleviate some of the pressure on these species. There is inadequate information regarding the fate of some of the smaller invertebrates of the Sundarbans. Further research into the diversity and viability of the remaining populations of many vertebrate species is imperative to establish future prospects for the faunal population of the Indian Sundarbans. Some threatened and extinct vertebrate groups are listed below:

THREATENED ANIMALS:

	Scientific name	**Common name**
Reptiles	*Crocodilus porosus*	Estuarine crocodile
	Varanus bengalensis	Bengal monitor

V. salvator	Water monitor
V. flavescens	Yellow monitor
Chelonia mydas	Green turtle
Eretmochelys imbricata	Hwaksbill turtle
Lepidochalys imbricata	Ridley's turtle
Caretta caretta	Loggerhead turtle
Dermochelys coriacea	Leatherback turtle
Lissemys punctata	Flap-shell turtle
Trionyx gangeticus	Ganges softshell turtle
T. hurun	Peacock softshell turtle
Batagur baska	Batagur turtle
Python molurus	Indian python turtle

LIZARDS AND SKINKS

The Geckos are a family of lizards of extraordinarily diverse form and is quite old in the evolutionary scale. Many geckos have a striking ability to change colour. They are active in the twilight and night. They occupy a great number of ecological niches.

Monitors (Varanidae) are one of the largest sizes. They are diarnal, reaching their full activity level when the sun is up and their habitats get warmed up. The slander, yellowish or brown olive Bengal monitor has numerous black spots. The two banded monitor, Varanus salvator is the second largest monitor dependent on water. The yellow monitor – V. flavescens is about a meter long.

Lizards/ Skinks

Hemidactylis brooki (House lizard)	*Chamaelon zeylanicus*
Hemidactylis flaviridis (Common house Lizard)	*Calotes versicolor* (Garden Lizard)
H. frenatees (House Gecko)	*Mabeya dissimillis* *M. carninata* (Common Skinks)

Varanus bengalensis (Bengal lizard) *Gecko gecko* (Wall lizard)
V. flavescens (yellow monitor lizard) *H. leschenaulti*
V. salvator (water monitor lizard).

Of all the lizards/skinks mentioned in the chart most conspicuous and common are large sized monitor lizards which are pride possession of Sundarbans. They are closely associated with all other fauna and interact with many of them. They are poached in large numbers for food and trade in skin.

8

Aquatic Avifaunal World of Sundarbans

Sundarbans estuary has more than 200 avian species. This is an approximate estimation. A checklist has been presented hereinafter.

[*Source:* Nandi & Mandal, ZSI]

The avian species may be broadly grouped into resident species, migratory species, species thriving shallow water, species of reclaimed areas and occasional visitors to the estuary. On a broad estimation the species distribution (conspicuous species) are as follows:

Resident	– 20 spp.	Migratory	– 16 sp.
Shallow water	– 5 app.	Reclaimed area	– 13 sp.
Occasional visitors	– 19 sp.		

Special importance of the avian species of the area is their breeding potential and the food habit. No work on the breeding potential of various species has been done. The work of Chaudhuri and Chakraborti on the breeding habit of three species of Cormorant birds, viz – Openbilled stork, Little Cormorant and Large egret has been recorded in this book (Sundarbans Mangrove – Ecology & Wildlife, 1987). But most interesting and useful study on the food habit of 23 species of birds done by Mukherjee has been recorded in the *Journal of Bombay Natural History Society* (1960^{s} & 1970^{s}).

SOME MARSHY BIRDS

Little grebe
Little cormorant
Oriental darter
Grey heron
Purple heron
Indian pond heron
Little heron.
Black crowned night heron
Chestnut bittern
Cattle egret
Large egret
Intermediate egret
Openbill stork

Pigmy cotton Teal
Eurasian moorhen
Eurasian coot
Bronze-winged jacana
Red wattled lapwing
Wood sandpiper
Little Stint
Whiskered Tern
Pied King fisher
White, breasted king fisher

Mookherjee of ZSI dissected stomachs of the aforesaid birds in various seasons and identified the gut contents both qualitatively and quantitatively. A short analysis of his work is reproduced hereinafter.

His work reveals how these birds are dependent on various species of invertebrate fauna (Mollusc, crustacea, etc.) and vertebrate fauna (lizards, reptiles, fish, mammals etc.) and on phyto and zoo plankters. The work clearly indicates that the aquatic and terrestrial habitats have to be meticulously conserved for the birds and all other groups of fauna to interact and survive.

A short analysis of his work (species specific food) will be very relevant as follows: (the list gives information on the varieties of food by each bird).

Little grebe

Crustacea
Vegetable matter
28.5% of diet
Annelid
Mollusc

Small fish 22.8%
Insects (Odonata and
Coleoptera)
Frogs 15.4%

Little Cormorant	
Genera of Fish	Fish 91.1%
Mystus, Anabas, Nandus, Chela, Puntius, Catla Oryzias etc.	
Oriental darter	
Genera of fish are Mystus, Amphibious Channa, Nandus, Mastocembelus, also Snakes and Lizards	Both fresh and brackish water species Fish 81.2% of diet
Grey Heron	
Genera (Fish)	
Mystus, Oryzias, Aplocheilus, Periophthalmus Boleopthalmus	Crustaceans - 24.8% of diet Genera Metapenacus, Scylla, Varuna, etc.
Purple Heron	
General (Fish)	
Mystus, Anguilla	Fish form 57% of diet
Periophthalmus, Boleophthalmus	Reptiles form 20.6% of diet
Crustacea 14.3%	
Little Heron	
Crustacea and Fish form 31.8 and 29% of diet respectively etc.	- Periophthalmus sp. Rana sp. (frog) and insects etc.
Black-Crowned Night Heron	
Frog: Rana Limnocharis comprised 27.9% of diet.	Lymnaea and Viviparus mollusc form 16.2% of diet
Fish, Crustacea, Annelids and insects form 12.2%, 6.3%, 5.9% and 10.4% of diet respectively	This is crepuscular and nocturnal bird

Chestnut Bittern	
Reptiles 13.4%	
Rana sp. (Frog) form 26.8% of diet	Others (the rest)
Insect 23.2%	Molluscs – Nerita, Pila comprised 11.6%
Cattle egret	**Some genera of insects are**
Attractomorpha	Dicladispa, Leptocorisa
Chrotogonus, Acryodium	Acrotylus, Hieroglyphus
Spodoptera	Insect form 69.8% of diet
Large Egret	**Common genera of fish are**
Fish form 80% of the diet	Mystus, Mugil, Aplocheilus Periopthalmus, Anguila, Polyemus & Ambasis Others - Mullusc, Shrimp, etc.
Openbill Stork	Snakes – Ptyas sp. and Xenochrophis sp.
Mollusc	
Pila, Lymnaea form 53.6% of the diet	Arthropods comprised 25.8% mostly Scylla serrata
Pigmy cotton Teal	
Vegetable matter 68%	Others
	Gastropods, Crustaceans, Lymnaea sp. preferred
Eurasian Moorhen	
90% Vegetable matter	Others – Mollusc
Eurasian Coot	
Others – Mollusc Insects Fish, Annelids	63.5% Vegetable
Bronze winged Jacana	
57.7% Vegetable Food	Others – Fish, Frogs, Insects

Red wattled Lapwing

Others – Gastropods   Insects 63.2%

Spiders etc.

AVIFAUNAL RESOURCES

To common public the avian fauna of Sundarbans are always an attraction. But few people have the field information that there are only a few selected sites where birds in interior Sundarbans aggregate or assemble for breeding purposes. Otherwise scores of kilometers cruise through the waterways of Sundarbans may not register sights even of a few species of birds. It is only in the fringe areas where the village swamps and wetlands exist, at least 30 to 40 species of water birds are commonly sighted; some may be in solitary pairs, others in large flocks. In the sea-face mud flats also one finds several species of inland and marine bird species.

Most exciting sights may be seen in late rains when most of the birds flock in hundreds to construct nests on low trees, lay eggs, rear fledglings and nurse them to the flight stage. One can register hundreds of prey species, which form the food of the fledglings and some predator birds, lizards and man.

This situation clearly indicates the inportance of swamps and wetland in the fringe area of the reserved forest. These swamps and wetlands are generally of fresh water though some are of brackish water also.

The fringe area, the mangrove forests and the coastal mudflats register more than 300 species of birds.

BIRDS OF SUNDARBANS

The mangroves, together with the extensive areas of mudflats, provide a wide spectrum of niches for avian species. Sajnakhali wildlife sanctuary is, in particular, an important bird sanctuary in the Sundarbans. More than 300 species have been recorded from the area, of which a number are migratory. At the onset of winter, in November and December, a large influx of birds of about 100 species visit the reclaimed areas from their breeding grounds in temperate regions further north. These birds leave the mangrove at the onset of the hot season, in March and April. Birds congregate in large numbers in both the reclaimed areas and mangrove forests. A list is presented showing a number of the more regularly observed species from the Indian Sunderbans.

Common birds of the Sundarbans include herons, cormorants, egrets, kingfishers, storks, darters etc. In particular, the openbill stork (Anastomus oscitans) is an important resident species and comprises more than 50 per cent of the total avifauna of the Sajankhali bird sanctuary, especially during the breeding season from June to September.

A number of prey birds such as eagles, including the White-bellied sea eagle, Crested serpent eagle, Palla's fishing eagle, Pariah kite, Shikra and owls such as the Barn owl, Scops owl, Spotted owl and the Brown fish owl are also found. On the inter-tidal mud flats, different species of Plovers and Sandpipers are common, while marsh birds such as Herons, Egrets, Bitterns, Storks and Rails are found throughout the year. Migratory flocks of Sandpipers, Redshanks and Whimbrels can be seen during winter. Ground dwelling birds, namely Swamp Partridge and Jungle Fowl are found in the dense forest.

CHECK LIST OF BIRDS OF SUNDARBANS

Grey Heron	Little Egret	Adjutant Stork
Purple Heron	Night Heron	Blacknecked stork
Large Egret	Little Bittern	Cotton Teal
Pond Heron	Chestnut Brittern	Pariah Kite
Little Green Heron	Black Bittern	Brahminy Kite
Cattle Egret	Openbilled Stork	Whitebellied Sea
Smaller Egret	White Ibis	Crested Serpent Eagle Gadwall
Pallas's Fishing Eagle	Indian River Tern	White-eyed Pochard
Shikra	Whitebreasted Water Hen	Brahminy Duck
Whitebacked Vulture	Purple Moorhen	Scaup Duck
Swamp Partridge	Moorhen	Lesser Whistling

Red Jungle Fowl	Eastern Skylark	Redcrested Pochard
Little Cormorant	Blackheaded Shrike	Kestrel
Green Pigeons	Blackheaded Oriole	Peregrine Falcon
Ring Dove	Black Drongo	Osprey
Spotted Dove	Wood Swallow	Marsh Harrier
Roseringed Parakeet	Ashy Swallow Shrike	Herring Gull
Koel	Starlings	Caspian Tern
Piedcrested Cuckoo	Greyheaded Myna	Common Tern
Brainfever bird	Pied Myna	Sooty Tern
Barn Owl	Common Myna	Gulibilled Tern
Scops Owl	Bank Myna	Whiskered Tern
Collared Scops Owl	Tree Pie	Yellowheaded Wagtail
Brown Fish Owl	House Crow	Paddy Field Pipit
Eagle Owl	Jungle Crow	Indian Tree pipit
Spotted Owlet	Redwhiskered Bulbul	Weaver bird
Jungle Nightijar	Redvented Bulbul	Greyheaded Lapwing
Palm Swift Indian roller	Jungle Babbler	Eastern Knot
Common Kingfisher	Old World Flycatcher	Little Stint
Indian Cuckoo	Paradise Flycatcher	Curlew Sandpiper
Crow pheasant	Mangrove Whistler	Little Ringed Plover
Blackcapped King-Fisher	Tailor Bird	Snipebilled
Whitebreasted King-Fisher	Blyth's Red Warbler	Dunlin
Whitecollared King-Fisher	Dusky Leaf Warbler	Turnstone
Lesser Pied King-Fisher	Chiffchaff	Whimbrel

Green Bee-eater	Paddy-field Warbar	Curlew
Hoopoe	Magpie Robin	Blacktailed Godwit
Coppersmith	Pied Bushchat	Common Red shank
Rufous Woodpeckers	Grey Tit	Wood Sandpiper
Lesser Goldenbacked Woodpeckers	Tickell's Flower Pecker	Terek Sandpiper
Larger Goldenbacked Woodpeckers	Purplerumped Sunbird	Green Sandpiper
Bronzewhinged	Purple Sunbird	Common Sandpiper
Jacana woodpeckers	White-eye	Fantail snipe
Pheasanttailed jacana	House Sparrow	Pintail snipe
Redwattled Lapwing	Baya	Green Shank
Little Ringed Plover	Whitebacked Munia	Rufoustailed Dove
Painted snipe	Spotted Munia	Wryneck
Little Tern	Giant Heron	Common Swallow
Blackbilled Tern	Spotted-billed Pelecan	Black Redstart
Indian Lesser crested Tern	Common Pochard	Collard Bush Chat
	Common Teal	Blue Rock Thrush
	Pintail	Blue Throat
		White Wagtail

9

Role of Honey Bee

Honey bees (Apis dorsata-rock bee) are active in this tigerland from Mid March to Mid June. The Maulies (Honey collectors) who go to the forest for collection of honey. Maximum number of fatalities (human death) occur in this tigerland, being the victims of man-eater tigers, during this period and the honey collectors are the maximum victims. A statistical figure is given below which will testify to the seriousness of the issue.

Casualty amongst professional men due to Tigers in the Sundarbans

Year	Honey Collector	Fishermen	Timber Traders	Others	Total
1964	10	13	3	15	41
1965	7	9	3	5	24
1966	9	20	2	15	46
1967	9	18	4	5	36
1968	13	13	2	13	41
1969	19	11	13	8	51
1970	11	12	6	3	32
1971	7	16	1	4	28
1972	13	9	-	-	22
1973	2	9	1	4	16

(contd.)

Year	Honey Collector	Fishermen	Timber Traders	Others	Total
1974	7	8	1	1	17
1975	14	40	-	12	66
1976	5	16	-	25	46
1977	11	30	-	6	47
1978	6	39	-	3	48
1979	11	41	-	1	53
1980	4	48	-	2	54
1981	3	23	1	1	28
1982	13	33	-	1	47
1983	5	15	1	-	21
1984	6	13	1	-	20
1985	5	27	-	-	32
1986	9	16	2	1	28
1987	7	26	-	-	33
1988	7	18	1	1	27
1989	1	8	3	3	15

HONEY AND WAX PRODUCTION FROM THE SUNDARBANS

Year	Honey (in Quintals)	Wax (in kg)
1977	161.44	52
1979	386.63	1,538
1980	407.97	2,019
1981	472.8	2,904
1982	405.61	2,360
1983	486.77	2,850
1991	148.29	880
2002	164.15	1363
2004	207.97	1282
2005	221.19	1100

ROLE OF VARIOUS TREE SPECIES AS HOST OF HONEY PLANTS

Excoecaria agallocha (3.9%)	*Sonneretia apetala* (5.3%)
Ceriops species (11%)	*Bruguiera* sp. (3.5%)
Rhizophora mucronata (10%)	*Xylocarpus granatum* (2.8%)
Heritiera sp. (9%)	*X.mecongensis* (1.9%)
Aegialitis rotundifolia (1%)	*Aegiceros corniculatum*
Hental	(0.5%)

During the period from 1973 and 2004 the forest of Sundarbans have undergone a sea of changes due to pilferage, refugee settlements and forestry practices. A fresh review of the situation on the aforesaid context is urgently necessary.

Some useful data are furnished below –

- Length of a comb varies from 80 cm to 120 cm.
- Width of a comb varies from 40 cm to 80 cm.
- 10 kg of comb yields 3 kg of honey
- 13 kg of comb yields 4.6 kg of honey
- 15 kg of comb yield 10 kg of honey
- 20 kg of comb yields 14 kg of honey.

Generally one tree has one honey comb, rarely 2 or more. Some *combs* even touch the forest floor. Most of denizens including tiger eat this delicious gift of nature.

Phenology and honey bee habitat

The peak periods of flowering for different mangrove species are as below and as depicted in Bar diagram-1.

20th March to 15th April	*Aegiceras corniculatum, Acanthus ilicifolius, Suaeda maritima* and *Sisuvium Portulacustrum*
31st March to 15th April	*Phoenix paludosa*
5th April to 20th April	*Ceriops species*
1st May to 20th May	*Avicennia spp.*
20th May to 5th June	*Excoecaria agallocha*
15th April to 5th May	*Sonneratia apetala*

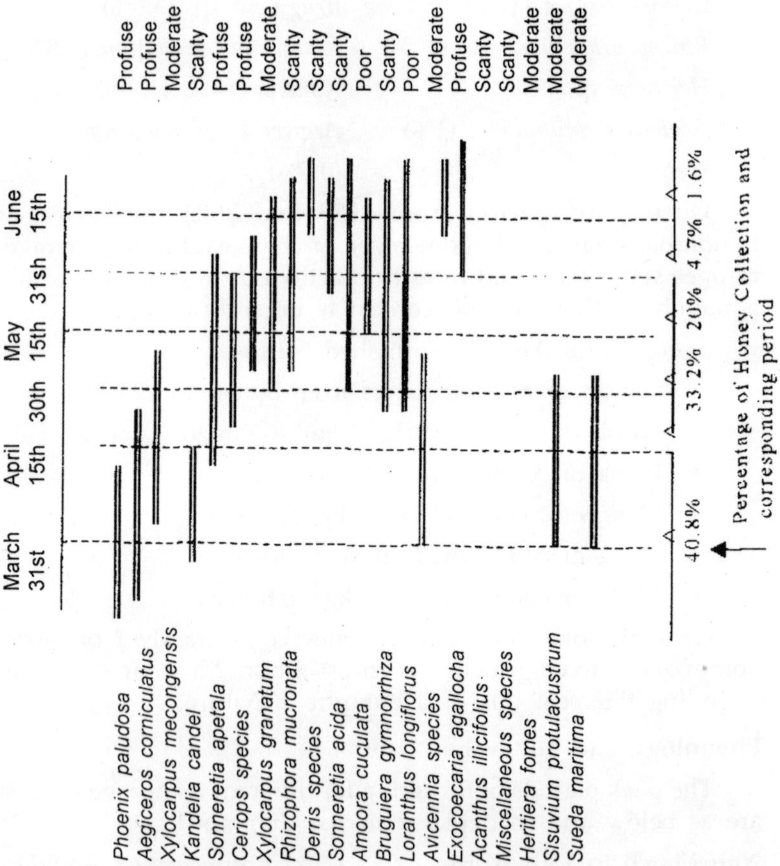

PHENOLOGY OF FLOWER PRODUCTION (PER CENTAGE OF HONEY COLLECTION)

Nipa fruitcans ═══ Scanty

Diagram 1

Honey formed from the first four species has thick consistency and is creamy white in colour. Honey made from *Aegiceras* nectar is considered to be the best and is cream coloured. Honey from *Ceriops* is reddish. Honey derived from *Sonneratia apetala* is yellowish and slightly light while honey from *Avicennia* spp. Has a reddish tinge and is light. The honey made from *Excoecaria agallocha* has a reddish colour and the taste is slightly acidic.

The colour of honey is affected by climatic conditions and the chemical composition of the nectar. The pollen grains of many species are coloured and are also reflected in the colour of the honey produced by the bees. The colour of pollen grains for some species is as follows:

Acanthus ilicifolius	yellow coloured pollen grain
Rhizophora mucronata	cream coloured grain
Bruguiera gymnorhiza	vermillion coloured
Xylocarpus species	yellow to deep brown grain
Ceriops species	cream coloured grain
Phoenix paludosa	red coloured grain.

Seasonality of honey production

Bulk collection of honey occurs in the following sequence from April to June

April 1st to April 15th	40.8 per cent.
April 16th to April 30th	33.2 per cent.
May 1st to May 15th	20.0 per cent.
May 16th to May 31st	4.4 per cent.
June 1st to June 15th	1.6 per cent.

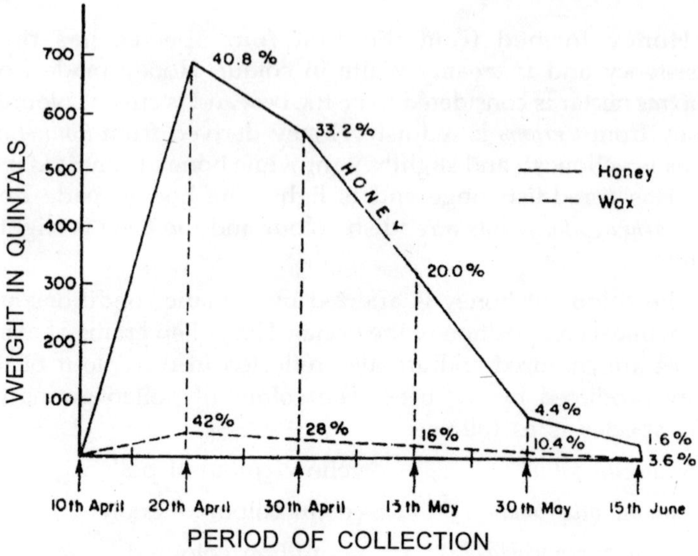

Yield of honey and wax between April and June

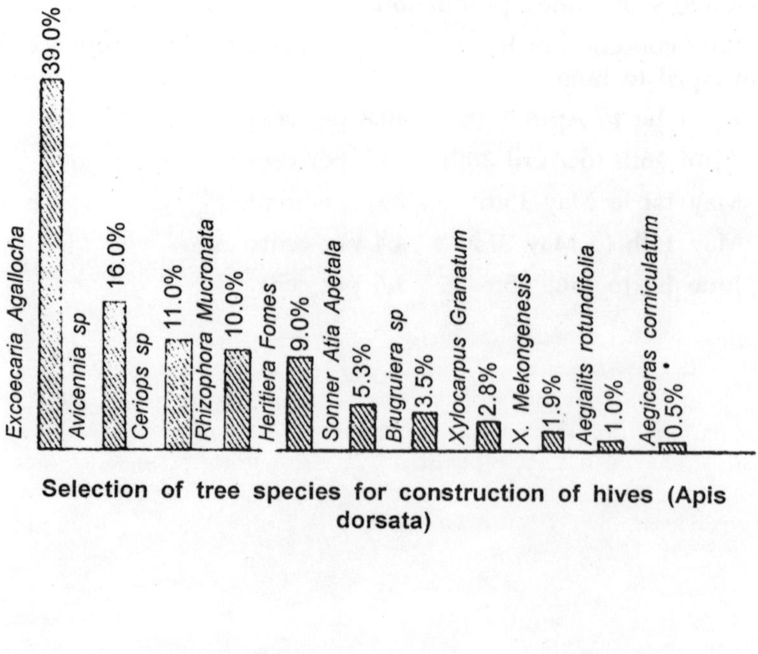

Selection of tree species for construction of hives (Apis dorsata)

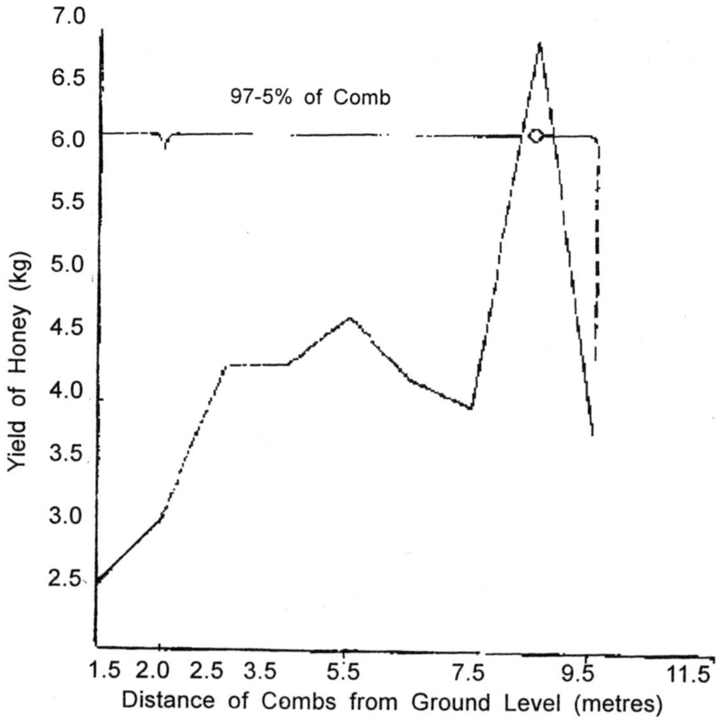

Relationship between honey yield and elevation of the hive

10

Ichthyofaunal (Fish) Diversity

The Estuarine water has both sweet (fresh) water as well as salt water fish. As such this region is rich in diverse species of fish. Very interesting species of fish are:

Tiger shark, Hammer headed shark, Saw fish, Electric rays, Eagle rays, Tenualosa ilisha, Lates calcarifer, Harpodon nehereus, Boleophthalmus and Periophthalmus spp. etc.

The fish catch is dwindling year after year. Hilsa catch is in decline. No restriction is imposed in catching of fish which are not allowable below a particular size. Mass scale capture of P. monodon went for years which damaged the various fish stock in the estuarine region.

State and Central Fisheries departments have done extensive ecobiological work on this subject.

FISH FAUNA OF SUNDARBANS

The estuarine water is very rich in ichthyofauna (fish). The following charts show the principal species of fish which are commercially exploited (in tones)

[*Source*: Jhingran & Jadav, 1989]

SALT WATER

Setipina sp (3109)	*Harpodon nehereus* (4216)
Tenulosa ilisa (1642)	*Polynemus paradiseus* (193)

Thichiurus sp. (2555)	Coilia *sp. (661)*
Pama pama (600)	*Sciaena biauritus*
Tachysurus jella (595)	*Lisa parsia* (22)
Ilisha elongata (443)	Others (7697)
Lates calcarifer (9)	Prawns (2118)

FRESH WATER

In the marginal area the following fresh water species primarily occur.

Macrobrachium rosenberjii	*Clupisoma garua*
Aorichthys aor	*Wallago attu*
Glossogobus giris	*Ailia coila*
Rhinomugal corsula	*Labio rohita,*
Eutropiichthys vacha	*L. calbasu*
Rita rita	*L. bata*
Catla catla	

Another yield figure shows the followings: (1993-'94 figures in tones)

Cat fish	–	975
Polinemids	–	276
Sciaenids	–	4,291
Bombay duck	–	5,276
Clipieds	–	10,181
Mullets	–	16
Ribbon fish	–	2,778
Prawn	–	3,979
Others	–	6,783

Other species commercially caught are:

Pangasius pangasius, Mystus gulio, Pampus argenteus, Mackerel, Liza tade, Sillago pajinus, etc.

GROWTH CHARACTERISTICS FOR HILSA ILISHA

The idea of determining k-value on growth characteristics of a few species of Sundarbans fish mooted out of some work published in the journal 'California Fish and Game'. This study was made in 1973 (statistical part was done by K. Chakraborti). It was one of the pioneer work done on Sundarbans fish.

(For detail refer the author's publications & bibliography).

Common fish species of the Sundarbans

Chillosctoma gasciatum
Carcharhinus limbatus
C. melanopterus
Glyphis gangeticus
Sphyrna blochii
Pristis microdon
Rhinobatos annandalei
Narcine brunnea
Dasyatis (Amphotistius) zagei
D. (Himantura) bleekeri
D. (Himantura) marginatus
D. (Himantura)uarnak
D. (Pastinachus) sephen
Aetobatus narinari
Corcia soborna
Gadusia chapra
Tenualosa (Hilsa) illisha
Anodontostoma chacunda
A. thailandiae
Gonialosa manmina
Nematalosa nasus
Ilisha elongata
Pellona dichela
Raconda russeliana
Coilia dussumieri
C. neglecta

M. cinereus
Pisocdonophis boro
Chanos chanos
Aorichthys aor
Mystus cavasius
M. gulio
M. vittatus
Silonia silonia
Pangasivs pangasius
Arius arius
A. caelatus
A. gagora
A. jella
A. platystomus
A. sona
Hemipimelodus jotius
Osteogeniosus militaris
Plotosus canius
Harpodon nehereus
Batrichthys grunniens
Rhynchorhamphus geogii
Zenarchopterus brachyopterus
Z. striga
Strongylura leiura
S. strongylura
Xenontodon cancila

C. ramcarati
C. raynaldi
Setipinna phasa
S. taty
Thryssa hamiltoni
T. purava
Megalops cyprinoides
Anguilla bengalensis
Moringua raitaborua
Lycondontis sathete
L. tile
Muraenesox bagio
Leiognathus blochii
L. equulus
Lutjanus argentimaculatus
L. johni
Datnioides quadrifasctatus
Gerres oyena
Gerreomorpha setifer
Pomadasys argenteus
P. nasta
Daysciaena albida
Johnius belangeri
J. coitor
J. sina
J. vogleri

Nibea soldado
Otolithoides biauritus
Pama pama
Drepane punctata
Oryzias melanostigma
Aplocheilus panchax
Monopterus cuchi
Lates calarifer
Chanda nama
Parambassis baculis
P. ranga
Terapon jarbura
Sillaginopsis panijus
Sillago sihama
Caranx carangus
Mehalaspis cordyla
Polynemus paradiseus
Apocryptes bato
Awaouichthys menoni
Boteophthalmus boddaerti
Brachygobius nanus
Glassogobius giuris
Gobiopterus chuno
Periophthalmodon schlosseri
Periophthalmus Koelreuteri
P. pearsei
Pseudaprocryptes lanceolatus
Scartelaos histophorus
Odontamblypus rubicundus
Amblyotrypauchen arcyocephalus
Trypauchen vagina
Kurtus indicus
Enpleurogrammus muticus
Lepturacanthus pantulai

Scatophagus algas	*L. savala*
Nandus nandus	*Trichiurus gangeticus*
Etroplus suratensis	*Pampus argenteus*
Liza microlepis	*P. chinensis*
L. parsia	*Stromateus sinensis*
L. subviridis	*Euryglossa pan*
L. tade	*Cynoglossus cynoglossus*
Mugil cephalus	*C. lingua*
Rhinomugil corsula	*Triacanthus brevirostris*
Valamugil cunnesius	*Chelonodon cutcutia*
V. speiglert	*C. patoca*
Eleutheronema tetradactylum	*Fugu oblongus*
Polydoctylus indicus	*Lagocephalus lunaris*

LENGTH WEIGHT RELATIONSHIP

Statistical analysis was carried out to find length weight relationship if any, and coefficients of condition were established for 4 economic fish species viz., Hilsa, Parse, Bhangan and Bhetki.

The following length weight relationships were obtained:

(1) $W = 166.0 \ (L)^5$ for Bhetki

(2) $W = .015 \ (L)^3$ for Parse

(3) $W = 0.46 \ (L)^2$ for Hilsa

Where W is weight in grammes and L is length in centimeter.

Hilsha (Hilsa ilisha)

Average length of fish catch varies between 24 to 48 centimetres while the weight varies between 125 grammes to 1950 grammes; the coefficient of condition is 1.29 at 29 cm. length while it is 1.93 at length 46 cm. which is maximum.

Bhetki (Lates calcarifer)

Average length varies betwen 20 cm to 97 cm, and the weight reaches a maximum of 4100 grams; the coefficient of condition is 2.55 at length between 38 to 46 cm; then the value falls off.

Bhangan **(Mugil tade)**

Average length varies between 15 cm to 62 cm., while the weight varies between 30 garms to 3000 grams; the co-efficient of condition is 2.74 (a maximum value) at length 28 cm.

Parse **(Mugil parsia)**

Average length varies between 12 cm. to 19 cms., while the weight varies between 20 grams to 110 grams; the coefficient of condition is 2.19 (a maximum value) attained at a length of 16 cm.

Significance of coefficient of condition:

The coefficient of condition is given by the formula:

$$K = \frac{10^5 \times W}{L^3},$$

where W is weight in grammes and L is the length in centimeter.

GROWTH

Growth of Organisms means three dimentional change in body and also of weight with increasing age. The change in organisms is due to conversion of food into body with body length and weight of an individual are plotted against specified terms here the graph comes out with a vector diagram known as growthcurve which appears S-shaped (sigmoid) and illustrates the rate of growth of the organism through time.

The rate of growth in fishes varies from species to species, with change of habitat, alteration of seasons or with the external factors of the environment influencing the physiological functions. The growth pattern and growth rates are highly species specific in fish. Under normal conditions the growth determining factors consist of (1) optimum temperature (2) adequate and correct food (3) presence of vitamins in food, particularly vitamin C whose deficiency may cause retarded growth and (4) periodicity of seasons of rapid and slow growth in the year.

Length-Weight relationship

Weight of fish is the function of its length. Theoretically, it is expressed by the formula (the cube law):

$W = KL^3$, where w = weight L = length, K = constant.

Since the cube formula does not hold good throughout the life history of the fish, as the value of K is not constant but subject to great variation. The most befitting formula is:

$W = L^n$ in which W = weight, L – length and n = constants.

This can also be expressed as:

$\text{Log } W = \log C + n \log L$

The values for the constants (a and n) are determined empirically from data, as the coefficient of condition.

Condition factor or Coefficient of condition

Condition factor is an expression of the condition in which the individuals fish, has been during certain period. Thus the mathematical expression of condition coefficient is a most effective and accurate method of evaluating the impact of environmental stresses on over all health condition of the affected fish.

Condition in general expresses the relative plumpness of fish with respect of the same species taken from different water bodies. It is expressed by relating length of fish to its weight. A fat fish will give a higher factor for condition than is given by a lean and thin fish. The methods of determining condition coefficient are as follows:

Coefficient of condition

$$K = \frac{\text{Weight in grams} \times 100}{\text{Standard length in cm}}$$

Condition factor

$$CF = \frac{\text{Weight in lb.} \times 100{,}000}{\text{Total length in cm}}$$

Index of Condition.

$$C = \frac{\text{Weight in lb.} \times 10{,}000}{\text{Total length in cm}}$$

The values of the factors will indicate greater plumpness while the lower values will reflect poor condition of the organism.

[Source : CIFRI]

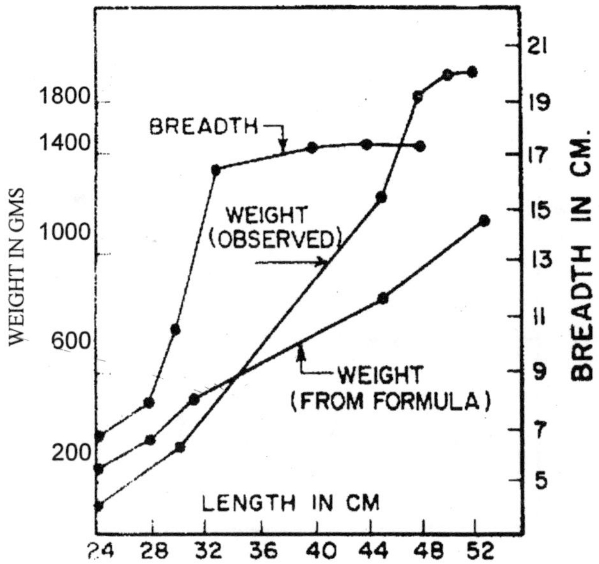
1800
1400
1000
600
200
WEIGHT IN GMS
BREADTH
WEIGHT
(OBSERVED)
WEIGHT
(FROM FORMULA)
LENGTH IN CM
24 28 32 36 40 44 48 52
21
19
17
15
13
11
9
7
5
BREADTH IN CM.

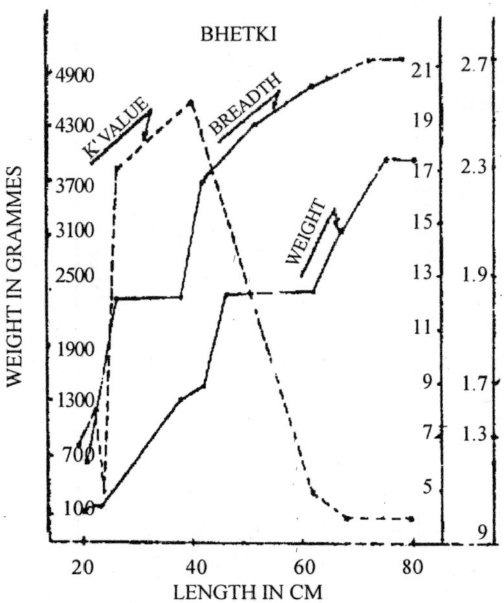
BHETKI
WEIGHT IN GRAMMES
4900
4300
3700
3100
2500
1900
1300
700
100
K' VALUE
BREADTH
WEIGHT
21
19
17
15
13
11
9
7
5
9
2.7
2.3
1.9
1.7
1.3
20 40 60 80
LENGTH IN CM

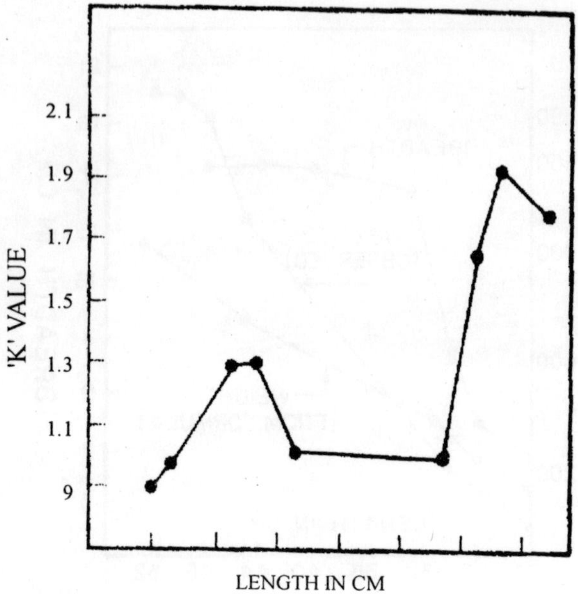

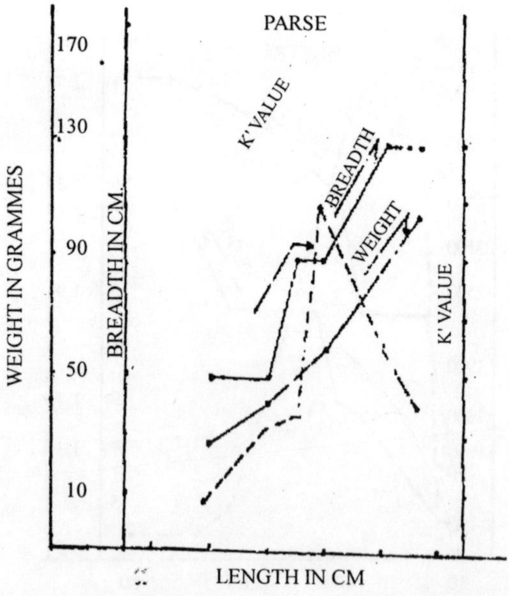

Growth characteristics of Hilsa, Bhetki and Parse

ESTUARINE AND COASTAL FISHERIES

The economic fish fauna of the Sundarbans estuary have been studied by Naidu (1952); Hora and Nair (1944); Pillay (1954); David (1956); Jhingran (1982) and Chakraborti et al. (1984). Surveys conducted in the Sundarbans indicate a total annual catch of fish of 25,000 metric tonnes. On an average, 4,000 individuals daily engage in fishing, with an average catch per fisherman of 1.5 kg of fish per day. From September to February, an average of 6,000 persons are normally engaged in fishing and this period accounts for almost 70 per cent of the annual fish harvest.

Shrimps and prawns constitute a commercially significant taxonomic group. The species include *Acetes* sp., *Macrobrachium lamarei, M. rosenbergi, Metapenaeus brevicornis, M. monoceros, Parapenaeopsis sculptilis, Penaeus indicus and P. monodon.*

There has been a sharp decline in fish catches from the Sundarbans estuaries in recent years. This has been caused by a combination of a number of factors. A steady increase in the number of fishermen has resulted in sharp competition, leading to over-exploitation. This, in turn, has led to a decline in numbers of adult fish reaching breeding condition, which are needed to replenish stocks. The fishermen also catch fingerlings and immature fish in their nets.

The increased rate of siltation, due to a lack of freshwater and excessive sediment load of waterways, has made rivers and creeks shallow, and fishing difficult. In addition, salinity levels have increased in all the rivers and creeks which is detrimental to fish breeding. Table A and B provide data on the volume catch for selected species in different years.

Overall production of fish and prawn from the estuarine Sundarbans has improved from 3203.9 tonnes in 1960-61 to 36,167.5 tonnes in 1994-95 which from 91.6% to 95.5% of total fish catch in Hooghly-Matla estuarine system (M. Sinha, CICFRI).

FISH SEED FOR AQUACULTURE

Huge quantities of fish seed are found in virtually all the river systems that are free of pollution. Prawn and fin fishes are caught by vast numbers of men, women and children using small nets. However, as collectors are only interested in *P. monodon* seeds because they fetch a higher price, great wastage of resources occurs through the indiscriminate killing of fin fish seeds.

Table 2: Dominant fish and prawns species in the estuarine area

Species	Fish catch (in tonnes)			
	No. 85- Oct. 86	Nov. 86- Oct. 87	Nov. 87- Oct. 88	Feb. 89- Jan. 90
Hilsa ilisha	2252.7	1935.3	1177.0	1563.4
Pama pama	3908.4	1090.8	3514.8	3978.2
Setipinna sp.	2055.8	1196.5	3852.5	4379.4
Trichiurus sp.	1473.3	1451.3	2502.3	3978.6
Harpodon nehereus	5197.7	2596.5	5084.1	4735.7
Tachysurus jella	702.2	366.6	758.0	432.8
Stromaleus cinereus	344.4	357.8	682.2	160.1
Polynemus paradiseus	106.5	141.4	179.4	98.2
Coila sp.	479.1	536.7	1591.5	537.2
Hilsa alongata	315.7	288.3	838.8	413.4
Sciaena biaduritus	602.3	92.6	604.4	87.5
Polynemus indicus	31.0	147.2	174.9	191.3
Prawns	1,892.4	2,563.5	1,787.0	2,670.0
Others	7,074.3	7,039.4	8,399.7	9,906.8
Total	26,435.6	19,797.9	31,146.6	33,139.6

A rational policy is required in order to ensure the sustained harvesting of the stocks of commercial fish from the Sundarbans. Restrictions on the size of nets use in fishing are required and only nets a lowing the permissible sizes for catch should be employed in fishing. Necessary replacement of the stock and banning fishing in breeding and spawning grounds is needed. Zoning and seasonal fishing should be introduced to enhance the capture of fish when the K-value is expected to be maximum.

Table 3: Yields of selected species in the Hooghly-Matla estuarine system

Species/species group	Mean: 1983-86 (kg/ha)	Mean: 1987-88 (kg/ha)
Hilsa ilisha	2.23	1.31
Other clupeids	3.20	7.07
Cat fishes	0.69	1.13
Polynemids	0.28	0.42
Sciaenids	3.85	4.58
Mullets	0.03	0.02
Ribbon fish	3.21	2.78
Bombay duck	4.95	5.65
Prawn	2.69	1.99
Miscellaneous	8.90	9.66
Total	30.3	34.61

Source: C.I.F.R.I., 1992-93

Technology alone cannot lead to the improvement of fish resources without scientific research into the biological and physiological aspects of fish species. The concept of rational fishing requires a sound scientific database in order to succeed. Statistical analysis of periodic sampling enumerations of zone/fish species frequency needs to be initiated immediately, to assess the optimum quantum of stochastic fish exploitation in various zones (rivers/creeks etc.) of the Sundarbans estuaries. The results of such analysis can then be used to formulate, as well as modify, fishery policies.

BRACKISH WATER FISHERIES

The brackish water and inundated lands further inland are extensively utilized for fisheries. There are more than 250,000 brackish water fisheries in south 24-Parganas District Brackish water species include *Catla catla, Cirrhinus mrigala, Labeo bata* and *L. rohita*. Commercially important fish and prawn species bred in brackish water fisheries include *Lates calcarifer, Liza parsia, L. tada, Macrobrachium rosenberghii, Mugil cephalus, Rinomugil corsola, Mystus qulio, Penaeus indicus, P. monodon, Metapenaeus monoceras* and *Tilapia mossambica*. These species are reared in places where the salinity of the river water reaches 20-35 μ mhos/ cm. Juvenile fish and larvae of *P. indicus, Metapeneus brevicornis* and *M. nonoceros* can be naturally stocked in the estuarine fisheries from seed contained a tidal waters.

Salinity is one of the main controlling factors in determining the productivity of brackish water farms. There is an inverse relationship between salinity and dissolved oxygen. Hence, brackish water fish ponds should be shallow in nature with a large surface area so that atmospheric oxygen can be easily dissolved in the water.

The Central Inland Fisheries Research Institute (CIFRI) at Kakdwip has developed a technique for the polyculture of fish and prawns by which a production of 2,600 kg/ha/year has been achieved. Hatcheries have been introduced and have a tremendous impact on aquaculture in the region. It is now possible to earn Rs. 20,000/ ha per annum (US$1,597) or more, through fish and prawn culture (Dwivedi, 1982). In comparison, there are two kinds of captive fisheries, namely bheries and bhasabade, which are amongst the oldest techniques in the Sundarbans region. The average yields by these methods vary from 700-1,000 kg/ha.

The Fishery Development Corporation, Government of West Bengal has developed on area of 200 ha for fisheries. Other areas in Jharkhali and Mousumi island have also been identified for development of bhery aquaculture. The International Fund for Agriculture Development (IFAD) has extended assistance through the World Bank to the Department of Planning and Development of Government of West Bengal National banks are also providing loans for developing this industry.

PADDY-CUM-PRAWN CULTURE

A National Committee set up by the Planning Commission on the development of backward areas, recognized that favourable conditions exist in the Sundarbans areas for the rearing of prawns and for brackish water pisciculture. However, prawn fisheries require heavy investments of money which the individual fisherman cannot afford. The committee therefore recommended that the Government should set up a corporation for the development of paddy-cum-prawn culture in the Sundarbans, and that small holdings between 2-5 ha could be set up for prawn culture.

However, a fact finding committee on paddy-cum-brackish water fish/prawn cultivation and related matters set up by the Government of West Bengal in 1984 recommended that paddy-cum fishery along the Calcutta sewage discharge zone around Bidyadhari could continue, while paddy-cum-fishery should be banned in other areas. Brackish water fisheries must not be permitted except in forest cleared areas and wasteland. They also recommended that there should be a complete ban of fisheries between embankments and the river bed, without prior written approval of a Standing Committee.

Effective control of irrigation embankments is to be ensured and no embankment is allowed to be altered without written permission of the Irrigation Department. The maximum area of a paddy-cum-fishery operation was set at 40 ha in order to check the growth of large fisheries. The issue or removal of license for a fishery should be through a district level committee. The Bidyadhari, the Matla, the Kalindi and the Raimangal rivers are to be declared protected in order to protect the fish resources. Fish seeds apart from prawns are not be collected.

Despite these directives, a number of problems still exist with regard to paddy-prawn culture in the area. The total paddy land converted into paddy-cum-fish cultivation in the 24-Parganas was 277,729.29 ha in January 1984. Such cultivation has caused lower yields of rice crop since good paddy lands are being utilized for pisciculture.

Extensive damage to the fish populations has continued unabated. Prawn seeds are collected while seed of other varieties of fish are thrown away and destroyed in the process. Hence, there is considerable damage caused to fish population which could otherwise feed the local market. An estimation shows, out of the total catch of seed from coastal waters, 30 per cent are prawn seeds, and the remaining consists of seed of other fish species.

Various social problems have also arisen due to paddy-prawn cultivation. Since bhery owners earn quick, high profits within a short space of time, the fishery owners have to depend upon organized criminal gangs for protection of their fisheries. The incidence of dacoity, murder, theft and rioting are on the increase and the causes are attributed in part to increased prawn cultivation. This is creating not only law and order problem but also degrading the social, cultural and moral standard of the villagers.

Profits from the fisheries are concentrated in the hands of a few bhery owners, middlemen, wholesalers or exporters and there are meagre returns to the poor villagers who are engaged in collecting seeds. Fishery owners may abandon the fisheries after one or two seasons and it is very difficult to reclaim these areas for the cultivation of paddy. Furthermore, the high profits earned from the fisheries are not ploughed back in the area but are siphoned off to the urban areas.

FRESH WATER AQUACULTURE

Indigenous fishes either cultivated or occurring naturally in the rice fields during monsoon include *Barbus* sp., *Oxygaster* sp., *Ambassis* sp., *Ambhypharynogodon mola, Colisa* sp., *Rasbora* sp., *Channal ophicephalus* sp., *Clarias betrachus, Heteropneuster fossilis, Pangasius pangasius, Mystus* sp., *Ompok* sp., *Wallgo attu, Anabas testudinus, Glossogobius qiuris, Notopterus* sp., *Ambhipnous cuschla, Channa marulius, Lates calcarifer, Channa punctatus and Bhadis bhadis,* among others.

In 1974-1975 the productivity of tank fisheries did not exceed 0.4 tonnes per ha. Presently, the production level for the area is 0.8 MT/ha, whereas the State average 10.5 MT/ha. The rapid development of paddy-cum-fisheries in the region has been mainly due to private entrepreneurship. About 33,000 ha of such areas are now in the Sunderbans. The productivity of prawns is at present about 250 kg/ha, mostly consisting of *P. monodon* and *P. indicus.* The salinity in such areas varies between 10-35 µ mhos/cm. The fin fishes are mullets, *Liza parsia* and *Lates calcalifer* and about 500 kg/ha are harvested per year. In low salinity areas (2.4 µ mhos/cm), Indian major carps are cultured. In some areas *Tilapia mossambica* is also cultured.

The benefit of paddy-cum-fisheries has reached small and marginal farmers who have raised their income from Rs. 300 to Rs. 2,000 per acre. To meet the rapidly growing demand for prawn and fish seeds, the rural people of the Sundarbans have become engaged in collecting *P. monodon* and *Labeo parsia* seeds with the help of home-made nylon nets. *P. monodon* larvae are selected and juvenile fish seeds are destroyed, thereby destroying the natural growth of other species of fish. About 20-50 per cent of the larvae are destroyed during handling and transporting.

PADDY-CUM-FISH CULTURE

The practice of alternating rice cultivation with fish culture and rice-cum-fish culture were initiated by farmers at a number of places in the Upper Sundarbans and the economic condition of rural villages has subsequently improved.

Paddy cultivation in reclaimed areas began in 1785 and the main crop is autumn/winter paddy. The average production of paddy from this area is only 1,400-2,000 kg/ha, under adequate rainfall and optimum climatic conditions and supported by adequate irrigation and drainage facilities. A good harvest can be obtained once in three years. The mono-cropped rice fields remain fallow from January to July if irrigation is lacking. Occasionally these fields are utilized for cultivation of seasonal crops such as sugar cane *(Saccharum officinatum)*, jute *(Corchorus capsularis)*, wheat *(Tritium aestivum)*, musur dal *(Cicer arietinum)*, mung beans *(Phaseccus mungo, P. aureus, Lens culinaris)* brinjal *(Solanum melongena)*, chilli *(Capsicum frutescens)*, cow pea *(Vigna sinensis)*, mustard *(Brassica campestris)*, sun-flower *(Helianthus annus)*, safflower *(Canthanus tinctoricus)*, okra *(Abelmoschos esculentus)*, potato *(Solanum tuberosum)* sweet potato *(Ipomoea batatus)* and tomato *(Lycopersicum esculentum)*.

Today there are better irrigation facilities and high-yielding rice varieties such as Jaya, Padma, Ratna Palmal, Bani, Lathisal, Mahan, Kamala, Masuri, Pankaj, IR 30, NC 1281, OC 1323, CR 1018, Prasad, Saket-4, IR 24 and other indigenous hybrid varieties of rice can be successfully grown in the lowlands. But constraints continue to exist. Hence, Ghosh *et al.* (1985) proposed a package of practices for increasing paddy and fish production through a rice-cum-fish farming system. The integrated farming can form a two-lier agriculture-cum-aquaculture system, which will not interfere with each other and will be able to optimize production from a unit area within a single year. Further research is required to improve production and farming techniques as well as to reduce destruction of fish seed.

11

Special Roles of Molluscs

MOLLUSCS

Molluscs belong to a large phylum of animal kingdom. They may be seashells of ocean's shores or snails from garden and ponds. Lovers of shellfish food are well aware of the deliciousness of clams and scallops. These are greatly diverse in structure and have many curious forms. They show a wide evolutionary diversity. There is great diversity of expression in the calcarious shells of molluscs.

The Bivalves are one of the largest and best known of the various classes of shelled molluscs. They are also a peculiar groups of animals.

Molluscs in the mangroves represent a transition between sea and land fauna and play a very important role in the decomposition of mangrove litters and food-chain in mangrove eco-system.

Tide, salinity and type of detritus exercise control over the distribution of Mollusc.

Following list indicates the types of molluscs according to their feeding behaviours:

Litter feeders	Suspended feeders

Herbivores and deposit Feeders	Carnivores
Algae and detritus feeders	Deposit feeders mud skippers

Conspicuous Mollusc (phylum) in Sundarbans

Class : Gastropoda

- Nerita sp. - Benthic
- Neritina sp. - "
- Littorina sp. - "
- Stenothyra sp. - "
- Assiminea sp. - "
- Cerithidea sp. - "
- Nassarius sp. - "
- Onchidium sp. - "

Class : Bivalvia

- Modiotus sp. - "
- Anadara sp. - "
- Crassostrea sp. - "
- Corbula sp. - "
- Bankia sp. - "
- Nausitora sp. - "

Class : Cephalopoda (squids and Octopuses)

- Octopus sp. Pelagic

PS. There are also phylum. 'Annelida' – Segmented worm, Britstle worm, Earth worm and Leeches etc.

Conspicuous Species

Littorina scabra	*Nerita articulata*
L. undulata	*Neritina violacea*
L. carinifera	*N. smithi*
L. melanostoma	*Gandetica miliacea*

Very conspicuous common species

Belamya bengalensis	Telescopium telescopium
Lymnea acuminata	Indoplanorbis exustus
Pila globosa	Melanoides tuberculatus
Vivipara bengalensis	Melanoides scabra

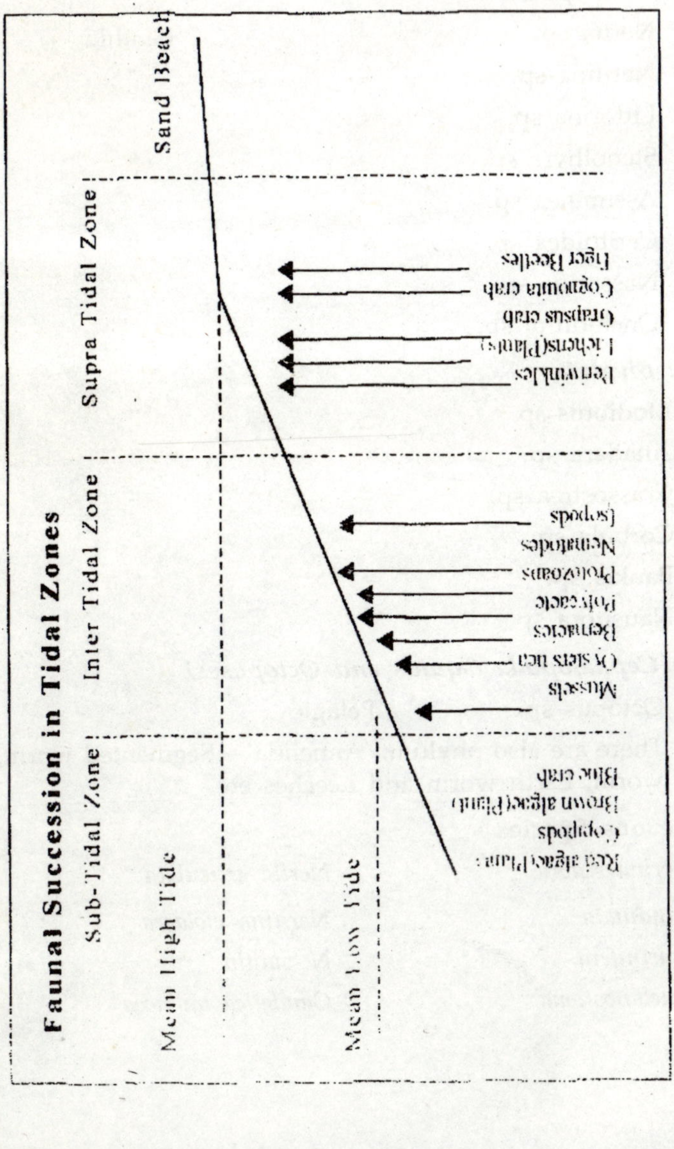

12

Special Roles of Crustacea

This group of invertebrates forms a very conspicuous, colourful and bio-ecologically very important faunal constituent in the estuarine forests of Sundarbans.

The tourists wild find an unlimited scope for behavioural study of these invertebrates in this terrain.

The subject being vast only a few genera and species have been mentioned in this abridged text.

Some Conspicuous species of Crustacea

Sand Crab	**Ocypoda (Pink coloured Crab)**
Metopograspus messer	*O. macrecira*
Sesarma bidens	
Macrophthalmus sp.	
Neosarnatium sp.	*Ceralophthalmus* sp.
**Macrobachium lamarei*	**Leander styliferus*
**Cardinia gracioipes*	**Metapanaeus brevicornis*
Varuna litterata (Graspid crab)	*M. monocoros*

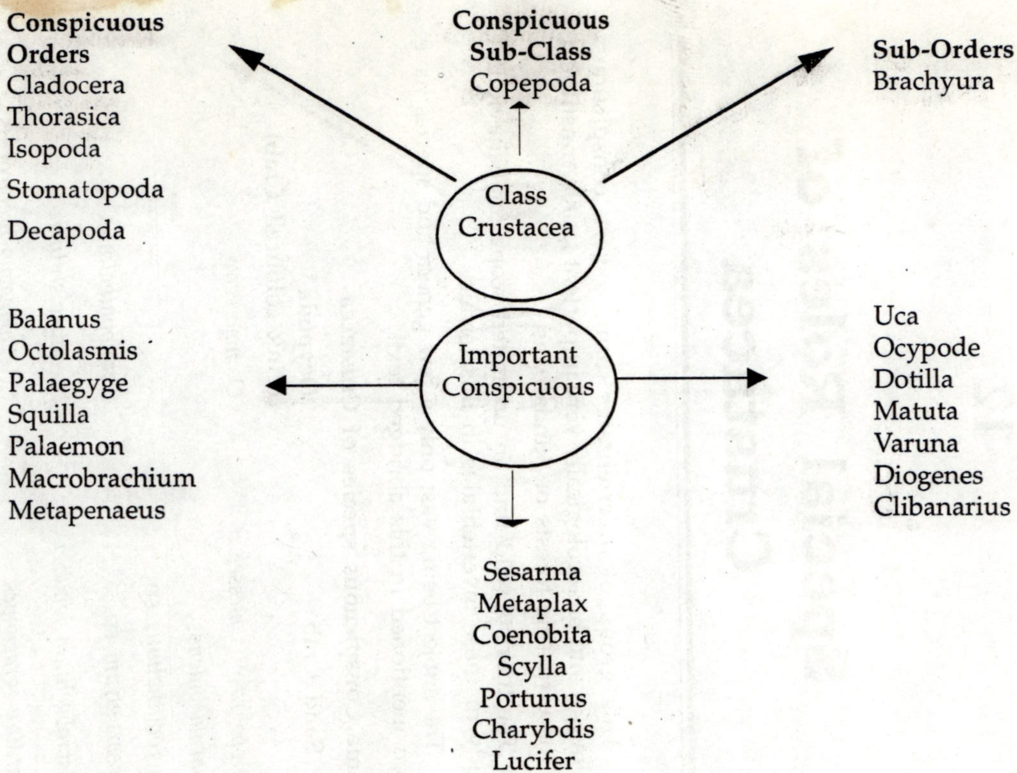
Conspicuous Orders
Cladocera
Thorasica
Isopoda
Stomatopoda
Decapoda
Conspicuous Sub-Class
Copepoda
Sub-Orders
Brachyura
Class Crustacea
Important Conspicuous
Balanus
Octolasmis
Palaegyge
Squilla
Palaemon
Macrobrachium
Metapenaeus
Uca
Ocypode
Dotilla
Matuta
Varuna
Diogenes
Clibanarius
Sesarma
Metaplax
Coenobita
Scylla
Portunus
Charybdis
Lucifer

Matuta victor

M. lunaris

Portunus sanguindentus

Dotilla blandifordii
(sand-buddler crab)

Scylla serrata

Portunus pelagicus

Limulus sp.
(Horse-shoe crab)

Colourful fiddler crab

Uca triangularis

U. acutus

U. annulipes

U. dussumieri

(* marked ones are freshwater crab)

NOTES ON A FEW GENERA

These crabs are wonderful creation of nature. They are colourful, very active on mud flats in low tide and display a variety of behaviour pattern.

This genus forms a large biomass among all crustacea. The crabs change colour with change of tidal level and is an ideal example of biological clock which is no notable among estuarine animals.

Cements with high sand content plays a key-role in the occurrence of *Uca lactacc, U. trangularis prefers muddy sediments. U. acuta* has wide choice of soil types.

There is seasonal fluctuation of species population. Some of the species are:

Uca dussumur, U. trangularis, U. acuta, U. lacteal.

HORSE-SHOE CRAB

This living fossil represented by *Corpius rotundicaudata, Limulus polyphenus, Tachypleus gigas, T. triclantadus, Xiphosura spolyphenus* in Sundarbans which are nature's rare gift. These crabs came to this earth 400 million years back during Upper Silurian age and are still found in Sundarbans forests.

Another species, the burrowing shrimp – "*Thalacina anomala*' goes down to 300 cm. below the ground is another nature's wonder.

Some special features are:

- Burrowing crab – *Ocypoda macrocera, Dotilla blandifordii*
- One sand – flat burrows made by bivalve *Sanguilaria* sp. And on mud flat by *Macoma birmanica*.
- Crawling traces made by *Nalica tigrina* (on sand) and *Cerithidea obtusa* (on muddy bed).

13

Benthic Biota

This chapter deals very broadly with benthic and planktonic flora and fauna which are very rich in Sundarbans. Extensive works have been done by Zoologists, Fishery and Plant Scientists on these items.

The author thinks that the nature hunters should appreciate and know that these flora and fauna as they play a very vital bioecological role in Sundarbans mangrove ecosystem.

BENTHIC FAUNA

These fauna live at the bottom of lakes, rivers and beels etc. where light cannot penetrate. They may burrow into the sediments and feed on them. These fauna indicate a good store-house of food for fish and other macro-fauna.

Benthic gastropod genera are – Cerithidia, Assiminea, Nerita, Littorina etc.

Benthic bivalve genera are – Dosenia, Meretrix, Modiolus, Acra, Solem, Theora, Pharella, Telina etc.

Benthic polychaete genera are – Diapatra, Spio, Marphysa, *Lumbrineris, Laoneoma, Glycera, Namalycastaris, Prionospsis* etc.

Benthic sea anemones are – Pelocoeles exul, Edwardsia jonessii, Phylocoeteopsis gangeticus, P. ramunni etc.

Benthic crabs are

- **Lower intertidal zone**

  Dotilla blandifordii, Scylla serrata, Uca dussumieri, Illyoplex gangetica, Metaplex crenulata, M. intermedia and Macrophthalmus pectinipes.

- **Mid intertidal zone**

  Uca acuta, U. triangularis, Metaplex indica, M. intermedia, Sesarma longipes,S. pictum, S. chiromantes, S. taeniolatum.

- **Intertidal littoral zone**

  Uca tringularis, Metaplex India, M. intermedia, Metapograpsus messor and M. maculates

Bioturbation structures result from activities of near shore benthic organism on sandy area. The animal groups are Anemones, Polychaetes, Gastropods, Decapods, Bivalves etc.

In muddy intertidal flats mounds are formed by *Macome birmanica, Uca* sp., *Cerithidia* sp., etc.

Benthic Flora

- Red algae - *Catenella* sp., *Bostrychia* sp., *Pterosiphonia* sp. *Caloglossa* sp. These occur on pneumatophores.
- On forest floor - *Vaucheria* sp., *Cladophorella* sp., *Oscillatoria* sp., *Lungbya* sp., *Aphanothece* sp.
- On mud flat (Diatom) - Species of Melosira, Navicula, Nitzschia, Pleurosigma etc.
- Sea-grass - Species of Zostera, Cymbodocea, Thalassia, Enhalus Holophila, Diplanthera etc.
- Other algae – Species of Spirulina Enteromorpha, Chaetomorpha, Gleocapsa, Gyrosigma Polysiphonia etc.

These flora indicate the richness of bottom biota. Water current, light, organic matter influence their production.

ZOOPLANKTERS

The zooplankter's potentiality is essentially alike in all tropical estuaries.

Most of the fishes are Zooplankton feeders. Some species starts as phytoplankton feeders which later switch over as zooplankton feeders.

Copepod group forms the primary food for molluscs, crustacea, fish etc. extracted to meet the need of poultry birds. The giant Ostrea colonies used to be exploited for lime production, which has since been discontinued.

Zooplankers belong to 13 taxa namely Copepods, Amphipods, Mysids Acetes, Chaetognaths, Polychaets, Lucifers Hydromedusae, Shrimp larvae, fin-fish larvae and Horse-shoe crab larvae etc.

Lucifers	Mysides
Gammarid	Copepods
Amphipod	Cumacea
Cladocera	Clenophores
Ostracod	

Copepods form the most important group in the entire zooplankton community 61–90%. Calanoida, Cyclopoida and Harpaclicoida have very high density (about 6000 numbers/m^3.

PHYTOPLANKTERS

Bloom formations are very significant feature of phytoplankton. In a same locality the bloom differ in size, shape, duration density and in biomass. The seasonal pattern of bloom formation, the biological rhythm of primary production, the sequential succession of species of phytoplankton determine the species interaction of the secondary producers.

The hours of sunshine, supply of organic nutrients, dissolved oxygen content in the water, alkaline pH, water current and circulation processes determine various aspects of bloom formation, the biological rhythm of primary production, the sequential succession of species of phytoplankton determine the species interaction of the secondary producers.

The hours of sunshine, supply of organic nutrients, dissolved oxygen content in the water, alkaline pH, water current and circulation processes determine various aspects of bloom formation.

Bloom forming genera are:

Biddulphia, Chaetoseros, Pleurosigma, Rhizosolenia, Coscinodiscus etc.

Algal bloom in shrimp form was a massive mass of Anabaena flosaquae.

Blooms impart colours to water like green dark, dirty brown, yellowish green, reddish, etc. Bloom may be formed by one or more species. Detailed list of algae is given below.

- Diatoms : Navicula, Pleurosigma, Gyrosigma, Cymbella, Cyclotella, Fragillaria and Amphipleura.
- Cyanophyceae : Scillatoria, Lynbya, Spirulina, Anabaena, Microcoleus, Nodularia.
- Chlorophyceae : Protococcus, Pediastrum, Hydrodictyon Ankistrodesmus, Nostoc, Volvox.
- Bacillariophyceae: Amphora, Nitzschia, Gyrosigma, Synedra, Gomphonema.
- Desmid : Cosmarium.

14

Faunal-floral Adaptations

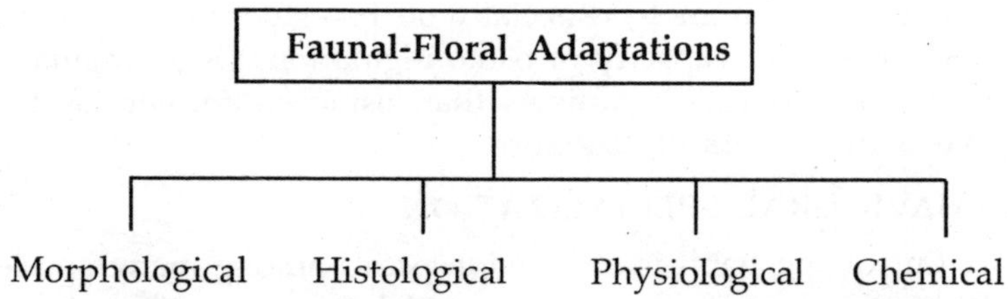

Researchers are still silent about any anatomical and physiological adaptations, developed among the man-eating tigers (primarily a terrestrial animal) as they survive by drinking saline water. This is to be unearthed if this practice has damaged the kidney or liver system, making them furious and compelling them to become man-eater.

The habitat conditions are varied, which comprise of water, water current, high tide and low tide, salinity, temperature variations, presence of various chemicals in water, reduced oxygen tension, more metabolic activities etc.

There may be anatomical specialization though scientists have not yet found any, but presence of salt-secreting glands associated with lacrymal glands have been recorded.

Flattened tails are the characteristics of some aquatic animals. Mudskippers have undergone structural modifications (fused pelvic fins), modified eyes of amphibians are few examples.

Crabs have shown greatest anatomical specialization related to feeding, respiration and vision (Grapsidae, Ocypodidae, Portunidae and Xanthidae). Development of strong jaw, chelac tips having spooned mechanism (Ocypode) feeding mechanisms of Dotilla and Uca are some instances.

Gill formation of some mud-flat fauna, strongly reticulated, pterigostomial region by raised ridges and setae which aid in spreading out water into a thin film over the reticulated region, helping in reoxygenation of water of the branchial cavity.

In Uca the gills are reduced, the gill chamber partitioned and many other adaptations are found in this genus. Ocular peduncle is the characteristic of most of the Ocypodids.

PHYSIOLOGICAL ADAPTATIONS

Osmo and ionic regulations are parts of many adaptations among several animals, especially of the estuarine crustaceans. Some have poor capacity of osmo-regulation. Osmo regulators are better estuarine organisms than osmo conformers as they have wide powers of tolerance.

BEHAVIOURAL SPECIALIZATION

Burrowing habit is most widespread among many species. The burrows are of many shapes stretching over various depths.

A complex feeding habits of tiger, monkey, wild-boar and other animals are also special adaptations to these animals of this estuarine environment, as they eat fish, snails, crabs etc.

Some estuarine animals possess the following characteristics:

- Power of habitat selection.
- Exhibition of biorythms in feeding behaviour.
- Active only at night.
- Special migratory capability. (Some decapod crustaceans)

Life history strategies of estuarine animals are interesting but very complex. Innumerable species display their own methods of adaptations to the dynamic habitat. It leaves wide scopes for the researchers to unearth the mystry of their behavioural adaptations.

[Modified from Fauna Conservation, series-3 (ZSI)-by AK Mandal & N.C. Nandi]

Micro-fauna of Sundarbans (At various niches)

Tree Community	Intertidal mudflat fauna
Estuarine fauna	Ground dwellers
Parasitic fauna	

Mangrove ecosystem—a very specialized one—supports a wide assemblage of animal communities. Most groups of animals live exclusively within the mangrove ecosystem. Some phylum and classes are as follows:

Phylum : Porifera, Clenophora, Platyhelminthes, Mollusca, Rhyncocoela, Ashelminthes, Chactognatta, Bryozoa, Annelida, Echiurida, Arthropoda, Cordata.

Class : Mammalia, Aves, Reptilia, Amphibia and Osteichthyes.

Macro-fauna of Sundarbans (Number of species in brackets)

Reptilia (57)	Amphibia (8)
Annelida (48)	Arthropoda (169)
Cordata (141)	Insecta (154)
Mollusca (94)	Mammals (40)
Gastropod (51)	Chiropetra (11), Primate (1)
Bivalve (51)	Rodenta (7), Cetacea (5)
Cephalopode (4)	Carnivora (12), Pholidota (1)
Arachnida (35)	Artiodactyla (2), Insectivora (1)

[Modified from Mandal & Nandi 1989 (ZSI)-F.C.S. No. 3]

Sundarban's faunal study should be made separately for forested and non-forested sectors (reclaimed areas). The above chart represents the fauna of the entire area. The forested area remains considerably undisturbed as compared to the non-forested areas, except for seasonal collection of honey, fuelwood and sporadic poaching of tiger, deer, pigs and lizards. The reclaimed areas have been largely brought under agriculture and fishery and hence these areas practically are devoid of existence of wildlife.

Along with felling of trees, now due to pilferage (formerly primarily because of large scale forest exploitation) the density of forest crop has been reduced considerably, besides extensive thinning out of tall trees like Sonneretia (keora), Xylocarpus (Passur) and Avicennia (Bain). As, such arboreal fauna has suffered considerably besides other fauna.

Large scale collection of shell, edible crab (especially Scylla serrata) and fish, the faunal (aquatic and amphibian) scenario has also considerably changed.

Littoral or supra-littoral forest fauna

A. Ground dwellers.

- Tiger
- Fishing cat
- Spotted deer
- Wild boar
- Frogs
  - Rana hexadactyla
  - R. crassa
  - R. tigerina and others
- Bandicoot rat
- Snakes:
  - Python
  - Coora
  - Krait
  - Viper, etc.
- Mongoose
- Otter
- Field mouse
- Grey mongoose
- Indian Crested porcupine
- Chinese pangolin
- Swamp Partridge
- Jungle fowl
- Burrowing lobster
- Gobicid fish
- Crabs:
  - Fiddler Crab (Uca sp.)
  - Sand-bubbler crab (Dolita sp.)

B. Tree dwellers.

- Rhesus monkey
- Bats:
  - *Pipistrellus* sp.
  - *Megaderma* sp.
  - *Cynopterus* sp.
  - *Pteropus* sp.
- Snails
  - *Cassidula* sp.
  - *Ellobium* sp.
  - *Littorina* sp.
  - *Nerita* sp.
- Snakes:

- Birds:
  - Parakeet, Dove Barbet, Roller, Kingfisher,
  – Crested eagle, white-sea eagle etc.
- Ants:
  - Sima rufouigra sp. etc.
- Slag:
  - *Onchidium* sp.
  - Indian bronze snake
  - Green whip
  - Woolf snake
- Honey bee
  - *Apis dorsata*
- Fish
  - Mud skippers

C. Intertidal mud flat

- Estuarine crocodile
- Gastropod
- Bivalve mollusc
- Decapod crab
- Polychaetes
- Crawling gastropods
  - Telescopium obtusa
  - Cerithidea sp.
  - Assiminea brevicula
  - Caenobita sp.
  - Assiminea brevicula
  - Caenobita sp.
  - Nerita sp.
- Plover
- Sand piper
- Dog faced water snake
- Eel fish
- Crab:
  - Sesarma sp.
  - Dotila sp.
  - Uca sp.
- Zooplankters
  - Nekton (fish sp.)
  - Bacteria
  - Protozoa
  - Mollusc

Notable Khal, Channel and Varani in Several Blocks

Netidhopani
Naubanki
Bhagwan-Bharani

Pirkhali
Kona
pich
Sudhanya

Panchmukhani
Gazikhali
Panchmukhani

Arbasi
Kalindi
Burhidabri

Matla
Lakhikali
Surajkali
Maraboni
Taratala
Storekhali

Choto hardi
Kachi
Narayantala
Bego

Gosaba
Mayadwip
Narayantala

Gona
Phulbari
Lodhidwani
Barkunda
Bagdalka

Mayadwip
Maldwip
Banka
Simri
Chotoarbhanga
Kedokhal

Chandkhali
Chakdkhali
Gorankhali
Boroduania
Bakultala
Khajurtala

Khatuajhuri
Natabar
Dakhin charra
Buridabri

Chamta
Chota Chamta
Bara Chamta

Jhilla
Chagmarikhal
Chilmarikhal

Bagmara
Dharkunda
Gorankhali
Barmara
Mechankhali

15

Sundarbans Biosphere Reserve (Aims and Objectives)

Notified in 1989, area: 9630 sq. km.

Forest: 4263 sq. km.

Conservation of Plant	*Conservation of Wildlife*
Maintenance of the balance of the ecosystem and the food chain.	Provision for intensive Research & Training for PA Management.
Introduction of Joint Forest Management.	Improvement of the socio-economic conditions of the fringe population, living below the poverty line.
Initiation of awareness Programme.	
Improvement of pisciculture.	Creation of Plantations.
Improvement of cultivation of honey.	Prevention of Soil erosion.
Creation of Self-help Groups	Problems of man-animal interaction. External influence in the area.

BIOSPHERE RESERVE

Under Man and Biosphere (MAB) programme of UNESCO (1970) Government of India declared this reserve a BR in 1989. There after this unique eco-system has been declared as World Heritage Site. This reserve has been included in the Network of Biosphere of UNESCO in 2001.

Biosphere reserve also caters to

- Enriching the existing flora and cover.
- Maintenance of the balance of eco-system & food chain.
- Obtaining Central Assistance for improvement of the reserve.
- World Bank Assistance for improvement.
- Spread of awareness programmes among the fringe people.
- Honey cultivation to boost Local economy.
- Self Help Groups.
- Pisciculture.
- Wildlife and their conservation.
- Minimisation of soil erosion.
- Important of the socio-economic condition of poor people.
- Man-animal interaction.
- Creation of forests.
- External influence in the area.
- Research.
- Joint Forest Management.

Other Biosphere Reserves of India

India possesses 11 such reserves which are:

1. Nilgiri BR. 5,520 sq. km.
2. Nokrek BR. 820 sq. km.
3. Nandadevi BR. 2,236.74 sq. km.
4. Manas BR. 2,837 sq. km.
5. Sundarbans BR. 9,630 sq. km.
6. Gulf of Manner 10,500 sq. km.
7. Great Nicobar BR. 885 sq. km.
8. Simlipal BR. 4,374 sq. km.
9. Dibru Saikhowa 765 sq. km.

10. Dehang-debang 5,111.5 sq. km.
11. Panchmari BR. 4,926.28 sq. km.

PLANS AND PROGRAMMES FOR IMPROVEMENT OF BIOSPHERE RESERVES

- Tree improvement research programme.
- Forest Resources survey.
- Creation of plantations of quick growing species.
- Forest consolidation.
- Creation of Mangrove plantations.
- Upgradation of management system.
- Strengthen wildlife species population.
- Improve Agro-silviculture.
- Monitoring and evaluation of forestry works.
- Creation and maintenance for Parks and Gardens.
- Improve Community Development works.
- Improve forest protection equipments.
- Increase amenities of forest staff.
- Economic rehabilitation of fringe people.
- Management and protection of forest and adjacent areas.
- Greening of urban and rural areas.

Of these sanctuaries and project Tiger areas Lothian Island is very close to Namkhana and the island is still in its initial formative phase. One can study various stages of plant and animal successions and associated issues. There is no risk of man-eaters, though care should be taken of man-eating estuarine Crocodiles. Bhagabatpur Crocodile and Turtle Research Complex is very close and this situation creates and ideal location for plant-animal studies.

Haliday island is very isolated and the animals there, particularly tiger have to negotiate large rivers for their survival. How can deer and wild boar population, obviously very small in numbers survive in such a small area even if there is a single tiger!

Sajnakhali WLS is a part of the project tiger area and is in the midst of a climax mangrove forest formation.

VISITORS OTHER ATTRACTIONS ARE: WILDLIFE SANCTUARIES

(In Mangrove Forests within Biosphere zone)

Biodiversity Protected Areas in Sundarbans

- Sajnakhali Wildlife Sanctuary (362.40 sq. km)
- Holiday Island Wildlife Sanctuary (5.95 sq. km)
- Lothian Island Wildlife Sanctuary (38 sq. km.)
- Sanjakhali Wildlife Sanctuary (1330 sq. km)

GEOGRAPHICAL INFORMATION SERVICE (GIS)

This cell was set up in Working Plan wing of West Bengal Forest Department in July 1999. It started working under A.K. Raha, IFS*, the then conservator-in-charge with one license PC Arc Info and Erdas software (Dos version). Subsequently two licenses of Arc Info GIS software, three licenses from ERDAS Image processing software, one Auto CAD map-2000, one license from Geomedia Professional and one license of Image Analyst with Micro station software have been procured for the GIS cell on proper authority.

The cell is presently manded by two departmental officers and three project personnel (GIS trained).

Some works undertaken are on:

- Forest Maps of the State.
- Procurement of satellite imageries from NRSA etc.

The Forest Department has a vision for the future as follows:

- Establishment of a few work-stations for field officers.

Generate forest fire maps.

(* Now Chief Conservator of Forests (South) & Director of BR.)

16

Sundarbans Estuary is a Vast Wetland

Areas below 6 metres depth of water as defined in Ramsar Convention, 1971, have been brought under the definition of wetland. This brings vast estuarine areas of Sundarbans waterways and land (due to tidal inundation) which covers practically all areas west of the river Matla and some to the east of the river and shallow rivers Saptamukhi, Thakuran, Ajmalmari, Dhulibhasani, parts of Matla and very many creeks and islands within the defined areas of wetland. So this is a vast wetland.

SUNDARBANS WETLANDS ARE A UNIQUE HABITAT

The Sundarbans swamps and forests stand supreme where diverse animal and plant species live in a complex physical, chemical, mechanical and animate environment. Here the plants and animals exist in perfect balance in a rich biomass by their stresses and strains, thrusts and counter thrusts. In this terrain the Tigers are amphibious and man-eaters and live in the midst of man-eating ferocious crocodiles, besides sharks, lizards, birds, porpoises, fish, turtles, honey bees, crabs, snails and other innumerable vertebrates and invertebrates.

The forests are halophytes that grow on swampy, clayey and muddy and saline substratum.

The estuarine tract of swamps water aid land form an

inexhaustible source of raw materials for human sustenance. Estuaries are considered as a "multiple user environment."

Observation, made in 150 A.D., corresponds very closely to the geographical features evident today. The main estuaries, from the West to east, are the Hooghly (Hugli), the Saptamukhi, the Thakuran, the Matla, the Bidya, the Ajmalmari, the Bidyadhari, the Gosaba, the Kalindi and the Raimangal. These rivers, apart from the Hooghly, have no connections with the head water channel of the upland river system.

Hooghly	The Hooghly river forms the western border of the Sundarbans and the South 24-Parganas district. The Hooghly is the main river of West Bengal and along with the Ganges and Bhagirathi rivers comprises a national waterway.
Muriganga	This river is a branch of the Hooghly river and flows along the east of Sagar island.
Saptamukhi	This river originates near Sultanpur and is connected with the Muriganga (Bartala) branch of the Hooghly river through the Ghughudanga Ganga (Hatania-Duania) and the Deogra Khal.
Thakurani	This river begins near Jaynagar in South 24-Parganas and has a number of connections with the Saptamukhi. It was perhaps earlier connected with the Calcutta Canal through the Kultali and the Piali rivers which exist today in a dying condition.
Matla	This river begins at the confluence of three rivers, these being the Bidyadhari, the Khuratya and the Rampura Khal, close to the town of Canning in South 24-Parganas. It was once considered the largest and deepest river of the area and was navigable throughout the year by ocean-going vessels. Canning was an important port connected with the Calcutta canal through

	the rivers Piali, Bidyadhari and Rampura Khal. The Matla is today connected with the Bidya and flows to the sea. There is no freshwater flow in the river any longer.
Bidyadhari	This was once a flourishing branch of the Bhagirathi but now serves only as a sewage and excess rain water outlet from the city of Calcutta. The river bed is completely silted and its connection with the Bhagirathi severed. The river remained active till the end of the 15th Century or early 16th Century.
Gosaba	The Gosaba is formed by the waters of Matla and Harinbhanga (Raimangal) though a large number of canals and flows through the reserve forest.
Harinbhanga	This river begins from Sahebkhali in North 24-Paraganas and is connected with the Rampura Khal by Barakalagachi river and with the Gosaba river through the river Terobanki. The Harinbhanga (also known as Ichamati and Raimangal) forms a natural demarcation between the border of India and Bangladesh.

The rivers of the region are tidal. Average tidal amplitudes in the estuaries vary from 3.5 to 5.0 metres. The highest amplitudes are recorded in July-August, while the lowest amplitudes are recorded in December-January. Free tidal flushing in the spill areas of these rivers has been interfered with by premature reclamation of their flood plains. As a result, large volumes of sediment carried by tidal water, which normally would raise the level of the delta face, are deposited on the river-beds because of the constricting effect of the embankments. These sediments have raised the level of the river-beds with respect to the surrounding land, thereby increasing the likelihood of floods.

The river flowing within the Sundarbans Project Tiger area may be divided into outer, middle and inner estuaries (Bandyopadhyay, 1986). The estuaries in Mayadwip and Gona blocks were classified as outer estuaries and those of Pirkhali and Jilla blocks as inner estuaries. It was found that salinity of the river water varies seasonally as well as with distance from the sea. As a result, salinity was higher in the outer estuaries than in the inner ones. Insignificant variations in salinity within the outer and middle estuaries were also recorded although a significant variation was found to exist in the inner estuaries.

A VAST WETLAND LANDSCAPE

If we accept the definition of wetlands as in Ramsar convention where marine water below 6 feet depth in low tide has been considered as wetland the vast area should be considered as wetland in Sundarbans. In a broad sense the rivers Saptamukhi, Thakuran, Ajamlmari, Dhulibhasani and Matla (Part) with all connecting channels comprising about 1500 sq.km comes under the area of wetlands.

This area has a vast fish and micro floral and faunal resources, besides birds, tiger, crocodile, spotted deer, pigs etc.

FISH AND FISHERY – WETLAND'S INVALUABLE RESOURCE

Fishes

The multitude of water bodies, including large and narrow rivers, channels and creeks that cris-cross the mangrove forests support many species of fish. The fallen leaves of mangrove plants enrich the soil-water phase and the environment creates and ideal nursery ground for innumerable species of fish.

The diverse and abundant population of fish species makes the taxonomic group of high economic importance. Consequently, considerable research has been undertaken in many aspects of fish biology in the Sundarbans. Notable works include those of Hamilton-Buchanan (1822); Day (1878, 1884), Annandale (1907, 1922); Hora (1933, 1934); Pantula (1965) and Shetty et al. (1961). On the other hand, ecological studies on fish and their interactions with various floral and faunal elements of the estuarine ecosystem is a vast field of study through research in this field is relatively new.

Two hundred and fifty species of fish have been recorded in the estuaries of the Indian Sundarbans, although most of the fish species are found during transient stages of their life cycles. They have a high degree of physiological adaptivity. Gobid fishes have deveioped certain morphological characters e.g., some lack eyes to adapt to living in the muddy substratum. Among the gobids, Boleopthalmus boddeerti, B. histophorus, Periopthalmus pearsei and P. schlosseri are common, while those found in the estuaries include Gibus chuno, Brachygobius nunus, Gobius alegans and Gobiades rudicaudus.

Among the eel species, the common species are Anguilla bicolor, Pisoodonophis boro and Muraenesox cinereus. Pangasius pangasius, a fresh water fish species, also breed in the estuaries. Arius, Mystus, Ostcogeniosus, Polydactyes, Plotosus canius, Liza tade, L. parsia, Lates calcarifer, Harpodon nehereus, Eleutheronema tetradoctrylum are other common fishes.

Some marine fishes that migrate upriver through the estuaries to breed in freshwater conditions include Hilsa, Pama, Sillaginopsis and Polynemus. Others only enter the estuary systems to breed, among which are Arius sp., Osteogeniosus sp., Polydoctylus sp. and Elutseronema sp. Freshwater forms which migrate to estuaries for breeding include the catfish, Pangasius sp. and eels such as Anquilla bicolor, Moringus raitaborua, Muraenesox cinereus and Pisodonophis boro.

Several species of sharks and rays also wander into the estuaries. Notable genera include Scoliodon, Pristis, Sphyrna, Rhinobatus, Dasyatus and Aecoatis. Sharks are common in some areas. A list of fish common species recorded in the Indian Sundarbans are presented.

17

Conservation Strategy

The Sundarban delta is an area rich in resource endowment but inhibited by poor people. Many of the resources of the region, namely, saline soil, brackish water, mangrove vegetation, mud flats, marshy lands animals and swamps, creeks, canals and rivers were often considered a waste. Attempts have been made in recent years very seriously to generate wealth out of these resources for the welfare of the human settlements and the well being of the environmental support system. This is an item which needs immediate attention of the policy planners and administrators of the State and the Central Governments.

One of the reasons for backwardness of and poverty in the region is the low literacy rate which varies between 41 and 55% for males and 13 and 25% for females. The rate is significantly low in respect of the Scheduled Castes and the Scheduled Tribes which constitute 42 per cent of the Sundarbans population. Increasing the literacy rate in the region should engage the immediate attention of the Government and the Panchayat Institutions. Similarly excessive addiction to drinks especially of the Scheduled Castes and Schedule Tribes population will have to be eradicated through intensive mass education programme.

There are lacunae in the system of land administration, utilization and reform in the region.

The jotdars and lotdars have reportedly cleared a substantial quantum of land for their occupation and use by evading the land ceiling laws.

The status report indicates that in all the sub-sectors, the productivity per unit area or unit effort is much lower than that prevailing in the State, or, for the matter in the country as a whole. In inland freshwater culture fisheries, per hectare production of fish is only 0.8 MT which is much below the State average of 1.5-2 MT of FFDA ponds. Similarly the productivity of bheri fisheries which is only 1 MT/ha is also low and requires substantial improvement. The yield of milch cows is similarly very low. This indicates that the Sundarbans delta merits receipt of special attention of Government to upgrade the present status of productivity in different sub-sectors to at least that of the rest of West Bengal.

The development efforts in agriculture sector and also in a few areas of non-agriculture sector have been analysed with special reference to indicators, namely, per hectare production, spread of high yielding varieties, land use pattern, rise in the level of income, contribution of GDP etc. Analysis of the development efforts indicated that the State Government as well as the very few voluntary organizations functioning the region have touched only the fringe of the problems of the delta. An indication may also be had of the totally inadequate investment made in the agriculture sector in the region. Similarly, NABARD's participation in the implementation of the different plans and programmes in the region has been of a very low level. There is no doubt that the quantum of financial investment and development efforts in the region should increase manifold to bring about rapid socio-economic development.

The analysis of the development efforts made so far indicates that only lop-sided attempts have been made to bring about socio-economic development.

The trend of reclaiming forest-land for rehabilitation has continued even after independence and such sustained attempts have been made only for agricultural purpose. This has done tremendous harm to the ecology and environment arising from shrinkage of forest areas, siltation of river beds and thereby, causing damage to Calcutta Port, and to the entire hinterland as a whole. The spill areas of Sundarban should, therefore, be left free for the tidal waters to play so that the beds of estuaries, rivers, creeks and canals are not silted up quickly. This is also necessary for regeneration of mangrove vegetation so very essential for the conservation of the ecology.

It is essential that suitable land use policy be chalked out immediately for Sundarban. Such a policy may be formulated only after a detailed investigation of the different resources available and the probable pattern of development in different sub-sectors in agriculture and other sectors.

One of the important constraints in the development of the region is the absence of an ecologically balanced effective progressive and a dynamic land use policy. Under the provision of the existing laws, fisheries is not considered as an agricultural activity.

An inventory of land resources should be prepared after a detailed Land and Soil survey. The National Land use and Wastelands Development Council have also recommended this issue in February, 1986.

The land use option indicated by the sample survey shows virtually equal weightage of agriculture and fisheries, even though investment and level of development in fisheries have been much lower than that in agriculture. Given proper extension., finance and technology support, fishery weightage in land use option is expected to increase significantly.

It is only through fisheries development the vast saline land and water expanses can be profitably used at low cost for economic development. Although the per capita income of the population is low, their consumption of calorie is comparatively high because of availability of fish.

In the low-lying areas, local aman paddy may continue in general but cultivation of summer and winter vegetables may be introduced in such areas by providing tank or canal irrigation. Supply of improved varieties of horticulture planting material may be given priority.

The percentage of double crop and multi-crop areas may be improved through provision of irrigation facilities. Such irrigation facilities will also encourage culture of fresh and brackish water fish. Along with paddy cultivation, cultivation of oil seeds, namely, mustard, sunflower, rapeseed may be encouraged.

Development of freshwater and mainly brackish water fisheries for generation of income, employment and wealth in the region should receive topmost priority. Increased production of fish and crustaceans from the region will also facilitate earning of valuable foreign exchange.

There is an urgent need for increasing the productivity level in freshwater culture fisheries from the existing level of about 8 MT per ha to 2-3 MT/ha through composite culture. There is also good potential for the culture of air breeding fish specially Clarius batrachus and Hetoropneustis fosilis since seed of such species are available.

Out of about 26000 ha under brackish water aquaculture in Sundarban, nearly 7000 ha are situated in lower Sundarbans which has a vast and unique potential for brackish water development in view of the abundant availability of inputs mainly, seed and physical and human resources, waste lands, mud-flats etc. particularly in the lower part of Kultali, Mathurapur-I and Patharpratima blocks. The present area of 26000 ha to about 50000 ha with the majority area coming under lower Sundarbans.

In culture fisheries, apart from the farming of Indian major carps, namely rohu, catla and mrigal, grass carp, silver carp and common carp, farming of mullets should be encouraged. This has been emphasized in the report of the National Commission on Agriculture, 1976. Under experimental condition, mullet culture has yielded as much as 2.4 tonnes per ha. Moreover, mullet seeds are abundantly available in Sundarban area.

Matla estuary is a unique capture fish, brackish water species, which have been affected due to pollution caused by the discharge of untreated sewage, siltation and stoppage of upland fresh water supply. An immediate survey should be conducted for the restoration of the health of Matla estuary in view of its tremendous importance from fisheries point of view in particular. Concrete steps should be taken by the Government for desiltation, restoration of upland discharge of water and control of pollution through legislative and administrative measures. If such actions are taken, the catch from this estuary alone may increase to about 10000 MT per year from the present figure of 3000-4000 MT.

Integrated fish farming, namely, paddy-cum-brackish water fish farming etc. may be encouraged in the areas, particularly up to the zones in Sandeshkhali, where Calcutta sewage influence exists.

FOREST

The existing forest areas should be conserved; some forest lands may be used for silvi-pisciculture which is immensely beneficial to the people and the environment. Pisciculture in the forest blanks may help regeneration of forests which is so very essential for the conservation of the ecology of the region. Silvi pisciculture programme in Dhanchi, Saptamukhi and Bhagabatpur forest zone is expected to yield additional production of prawn and fin fish of about 700 MT per annum. Such programmes should be gradually implemented in other forest belts which will help the section of the community connected with forestry and fishery sectors with high socio-economic benefits and employment opportunities.

Ecology and Environment

(a) The importance of Sundarban Delta in the conservation of ecology and environment of entire South Bengal should be properly recognized.

No further reclamation of the areas of the delta should be allowed for agriculture purpose only. Since this will result in the siltation of river-beds causing incalculable harm to Calcutta Port as well as its hinterland damaging, thereby, the entire environmental support system. If for inevitable reasons any embankment is required to be raised on rivers, creeks or canals, there should be sufficient spill areas for the river to play so that the bed does not get silted up quickly. "Drainage and reclamation of wetlands for agriculture, industrial development and other purposes can severely impair biological diversity, fisheries and environmental services"[1].

(b) Mangrove swamps should be conserved as they are essential for preventing erosion of coastline and other land and serve as shelters to the young fish crustaceans.

1. Ledee George and Goodand Robert, *Wildlands: Their Protection and Management in Economic Development*, World Bank (1988).

A detailed investigation should be made by the Government to identify the areas where spawning of fish and prawn takes place and such areas should be declared as fish sanctuaries beyond the reach of fishermen and others. Sanctuaries if set up will also act as areas for conservation of bio-diversity. Discharge of untreated sewage to the river system of Sundarban should be controlled immediately. Similarly, supply of upland discharge of fresh water into the river system should be ensured.

The modern trend in management of waste is its recycling and not disposal. Establishment of sewage treatment plants is expensive. Sewage-fed fisheries, which are natural biological reactors for converting waste into wealth should be encouraged for treating Calcutta sewage for its discharge into Matla. This will increase fish production, income and employment and reduce environmental degeneration.

Certain areas to be identified for management as wildlands i.e. direct maintenance, conservation and enhancement of virtually unmodified ecosystems and their native flora and fauna so that these areas give the greatest sustainable benefit to the present generation while retaining the potential for suiting the needs of the future. The management of wildlands in Sundarban is vital for preservation of biological diversity and environmental services like breakdown of pollutants provision of nurseries for the commercially important fish species and barriers against cyclone damage. Animal Husbandry. (Modified from B.C. Sharma's book, "Lower Sundarbans" 1994).

CONSERVATION POLICIES

A number of Acts and policies, in conjunction with the extensive network of sanctuaries, national parks, biosphere reserves and other protected areas, provide a substantial legal and policy framework for the protection of forests like the Sundarbans. There are three main laws for the protection of the environment. These are the Wildlife Protection Act (1972), the Forest Conservation Act (1980) and the Environment Protection Act (1986), with their various amendments (Government of India, 1992a).

The National Conservation Strategy and Policy Statement on Environment and Development (Government of India, 1992b) provides directives for the integration and internalization of environmental issues in the policies and programmes of various sectors. The National Forest Policy (1988) complements the Forest Conservation Act (1980) by encouraging people's participation in forest conservation activities and by promoting usufruct sharing of forest produce.

A National Mangrove Committee, comprised of experts and representatives of concerned government departments, was established in 1987. The composition of the Committee was reconstituted in 1990 and its task was defined as that of an advisory committee for the conservation, research and training requirements for mangrove ecosystems. In 1992, this Committee was combined with the National Wetland Committee to form a single National Committee on Wetlands, Mangroves and Coral Reefs.

Steering committees at the State level have also been established and are generally headed by the Chief Secretary of the State with the Secretary, Department of Environment acting as Member-Secretary.

The main problems identified by the committee are reflective of the issues described above and at greater length in the preceding chapters, and are as follows:

1. Exploitation of the mangrove forests due to human pressure.
2. Conservation of mangrove areas for different agricultural purposes.
3. Dwindling of fisheries.
4. Loss of flora and fauna, including endangered species (Government of India, 1990).

ACTION PLAN FOR THE SUNDARBANS

Under these Committees, an Action Plan for the conservation of the Sundarbans mangroves has been developed. Across the country Action Plans for the mangrove sites identified by the Committee comprise the following basic components:

- Survey and Demarcation of mangrove areas selected for conservation.
- Natural Regeneration Areas to be sheltered from anthropogenic stress

- Afforestation in severely degraded areas with the cooperation of the Forest Department and village councils.
- Nurseries to be developed near the afforestation site to provide necessary saplings.
- Protection, watch and ward programme to protect the mangrove area from encroachment and exploitation.
- Education and awareness programmes for mangrove dependent communities.

Research projects for the assessment of the impact of mass collection of prawn seeds and the survey, identification and screening of potential foliar fungi have been supported by the Ministry of Environment and Forests, in addition to continued efforts at regeneration of mangroves in degraded areas (Government of India, 1993-94).

These policy developments and activities indicate a growing awareness amongst researchers and policy makers of the importance of mangroves and the multi-faceted consequences of their destruction. However, information regarding forest cover and condition remains hazy and public awareness programmes remain weak. Poor social and economic development of the Sundarbans region continues to exert enormous pressure on the land and other resources of the ecosystem. As a result, despite efforts at natural regeneration and afforestation, monitoring activities and ensuring their long-term viability is, as yet, a daunting task. Public involvement therefore, is an essential component of the forest protection and improvement activities in the Sundarbans.

The mangrove forest of the uncleared Sundarbans forest of Bangladesh and India act as a coastal buffer, dampening the effects of storm waves, which can exceed four metres in height, and reducing the severity of gales. Mangrove ecosystems play an essential role in stabilizing the coastline and providing essential spawning and breeding areas for numerous species of commercially important biota. Once these protective and productive values are fully recognized by the public and private interest groups, the need to maintain and intensify efforts to protect and rehabilitate the mangroves of the Sundarbans will become self-evident.

LAND-USE POLICIES

The most important constraint to the development of the region is the absence of an effective, ecologically based, land-use policy. The existing laws do not consider fisheries as an agricultural activity and agricultural lands may not be used for fish farming. In 1976 the National Commission on Agriculture noted that the use of land should be optimized by observing the ecology, economy and capability of the area. A detailed land-use plan is necessary for this deltaic area. The National Land Use and Wastelands Development Council recommended in 1986 that an Inventory of land resources should be prepared. A follow-up to these programmes is necessary.

A sizable portion of Sundarbans have been reclaimed for human settlements and related activities over the past 200 years. As a result of poor, unplanned development, the reclaimed areas of the Sundarbans are, however, lacking in basic facilities and the people of the area live at a subsistence level. Studies have shown that the full potential of the area in terms of aquaculture development has not been realized or adequately addressed in previous development planning for the area. The role of the Sundarbans mangroves and the importance of maintaining a healthy relationship between the inhabitants of the reclaimed areas and the mangroves, must therefore be addressed in future development plans for the region.

CONSERVATION OF BIOLOGICAL RESOURCES

The high population density in the State of West Bengal, which includes one of the world's most populous cities, Calcutta, situated close to the Sundarbans, exerts immense human pressure on the resources of the area. A large number of faunal and floral species have vanished over the years as a consequence. The dire situation has forced the Government to establish several sanctuaries in the mangrove forest area.

The biosphere reserve in the Sundarbans includes a Tiger reserve under the Project Tiger Programme. The core area of Project Tiger has been converted into a national park which gives added conservation input and effort towards the protection of the biota. Another sanctuary at Sajnakhali has been created for the protection of resident and migratory birds, as well as the resident animals. Outside the tiger reserve, to the west, lies Lothian Island Wildlife Sanctuary, which is an ideal site for the study of plant and animal succession. Haliday Island is another wildlife sanctuary in the Sundarbans.

Alarmed by the steep decline in the world's tiger population, the International Union for the Conservation of Nature (IUCN)- now known as the World Conservation Union – declared the tiger (Panthera tigris tigris) an endangered species in 1959 and listed it in the Red Data Books. In India, the killing of tigers was prohibited in 1970, followed by a ban on all hunting.

A country-wide tiger census in 1972 revealed that the species count was down to less than 2,000 tigers, an alarming decline from an estimated 40,000 tigers at the tune of the present century (Panwar, 1991). Following the passage of the Wildlife Protection Act in the same year, tiger protection was taken seriously and on April 1, 1973, Project Tiger was launched in an initial nine reserves, covering a total area of 13,723 km^2 of which 4,242 km^2 were core areas. Presently there are 28 tiger reserves in 19 states across the country, covering a total area of 37,721 km^2 areas.

Tiger Project is a comprehensive programme with an aim to protect the natural ecosystem by mitigating human-induced limiting factors and to bring it as close to a natural functioning area as possible. Further emphasis is laid on the standard of living of the local communities living in close proximity to the ecosystem to enhance sustainable development in harmony with the carrying capacity of the natural habitat.

ECOLOGICAL REHABILITATION OF THE SUNDARBANS UNDER THE INTEGRATED WASTELAND PROJECT.

This project envisages a total economic uplift of local communities presently living in extreme poverty in remote areas of the Sundarbans. The project is being implemented through the Sundarbans Biosphere Reserve. The scheme has been designed to combine a full spectrum of activities, comprising afforestation, conservation of fragile areas, development of pasture, soil conservation, minor irrigation, cottage industries, and other socio-economic and ecological components.

The aim is to involve scientists, technologists, administrators, peoples representatives, NGOs, public institutions and fringe groups through participatory management techniques in a fully integrated programme. The project covers four years from 1991-92 to 1994-95 of the Eighth Plan period and includes the creation of 3,270 ha of plantations, soil conservation work on 133 km along stretches of embankments and 3,000 ha of natural regeneration plots. The scheme is expected to generate approximately 94,000 jobs over a period of four years.

PART II

FLORA AND FAUNA OF ANDAMAN AND NICOBAR MANGROVES RESOURCES, CONSERVATION STRATEGY, PROBLEMS, REMEDY

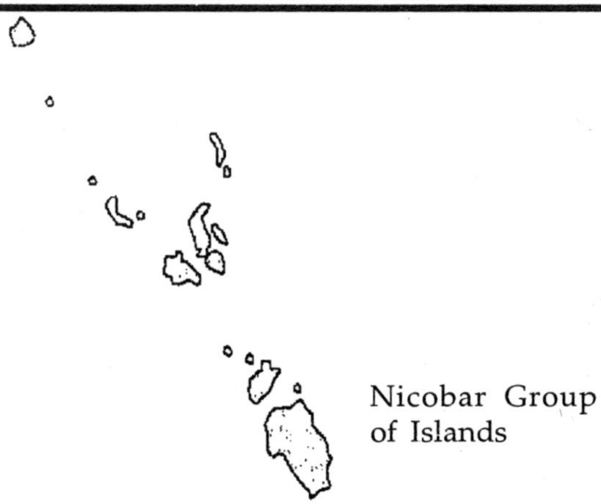

BOTH THE MANGROVES ARE INTERLINKED BY LARGE MARINE ECOSYSTEM (LMES) OF FAO

SUNDARBANS MANGROVES

- as a bridge between Terrestrial and Marine ecosystem.

It has
- Tiger Reserve
- Wildlife Sanctuaries
- Biosphere Reserve
- World heritage site.

↓

MARINE ECOSYSTEM

The Bay of Bengal has been identified as one of the World's 49 Large Marine Ecosystem (LMES). FAO executed Bay of Bengal Programme (BOBP) for development in 2002.

Bay of Bengal Marine Eco-System.

↓

Andaman Group of Islands.

ANDAMAN AND NICOBAR MANGROVES

Mangrove association

Has –
- Coral ecosystem
- Marine ecosystem
- Tropical Rain Forest Ecosystem.
- Influence of SE Asia's flora/ fauna
- Biosphere Reserve
- 100 protected areas
- Decaying tribes.

↑

Bay of Bengal

↓

Nicobar groups of Islands

Some endemic birds of Andaman and Nicobar Islands
(Photos: Courtesy of 2.5,1)

Great Nicobar Serpent Eagle (*Spiloronis klossi*)

An Endemic Species Limited to Great Nicobar Island

Narcondom Hornbill (*Rhyticeros narcondamsi*)

Endemic to only Narcondam Island

Andaman Crested Hawk Eagle (*Spizaetns Cirrhatus* Andamanensis)

Endemic to Andaman Group of Island very common in mangroves

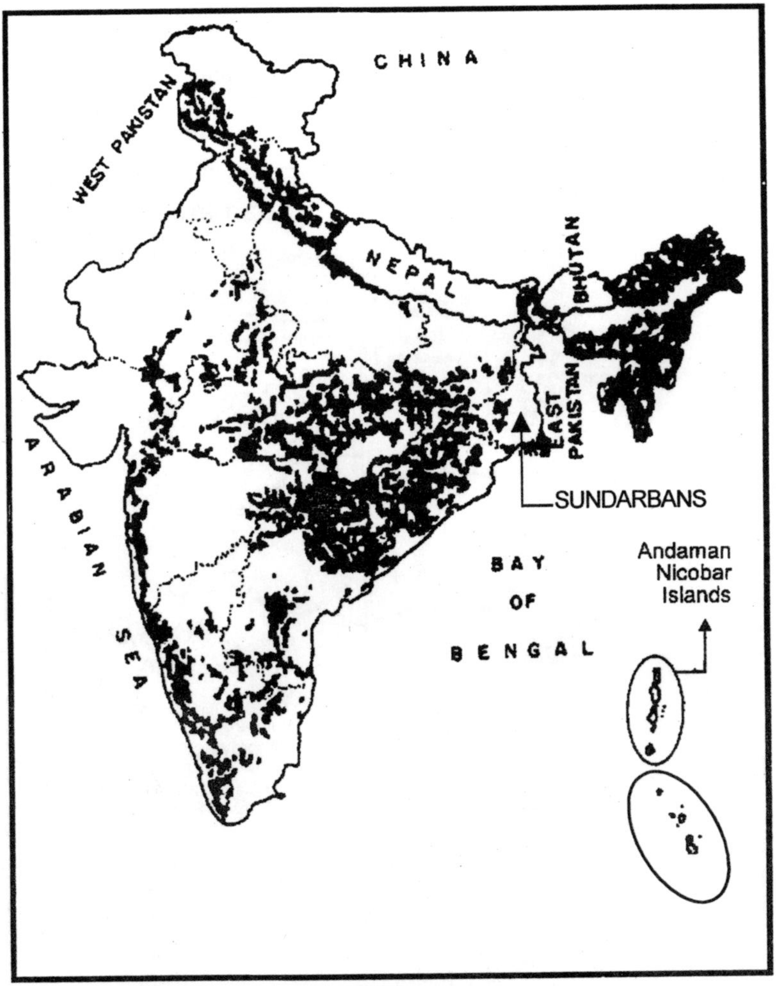

Existence of forest areas in mid-twentieth century
After 50 Yrs the forest

Map 1. Andaman

Diversity of aminals, birds and forests of Andaman and Nicobar islands seeing is thousands time better than hearing and reading. Hence efforts have been made to depict the salient features to forests in maps to get an idea at a glance.

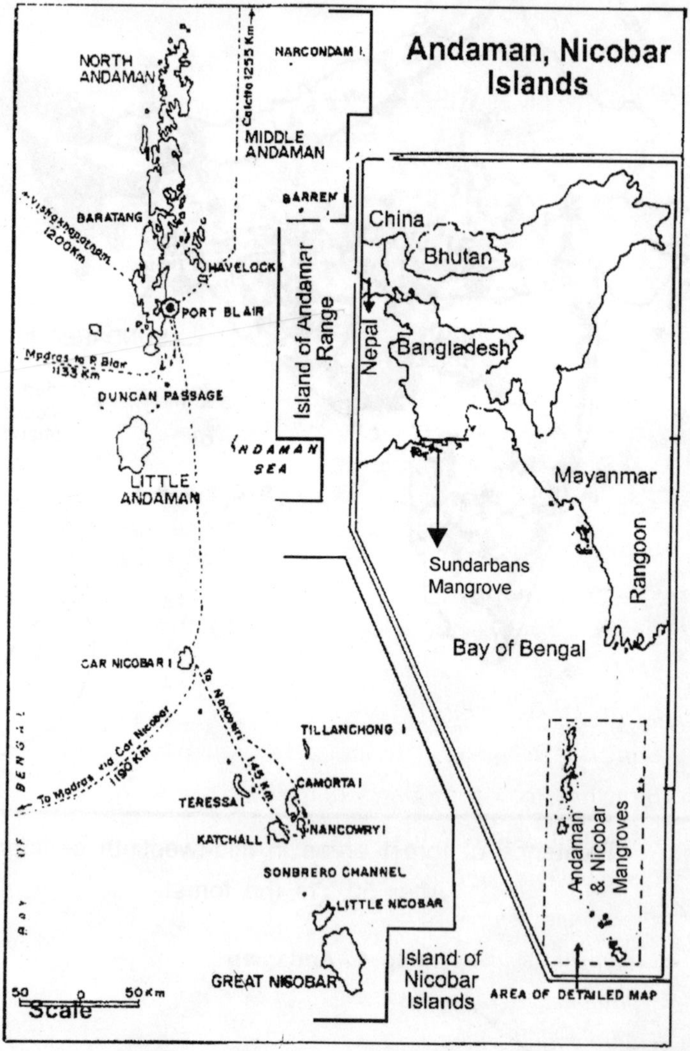

Map 2. Andaman and Nicobar Island

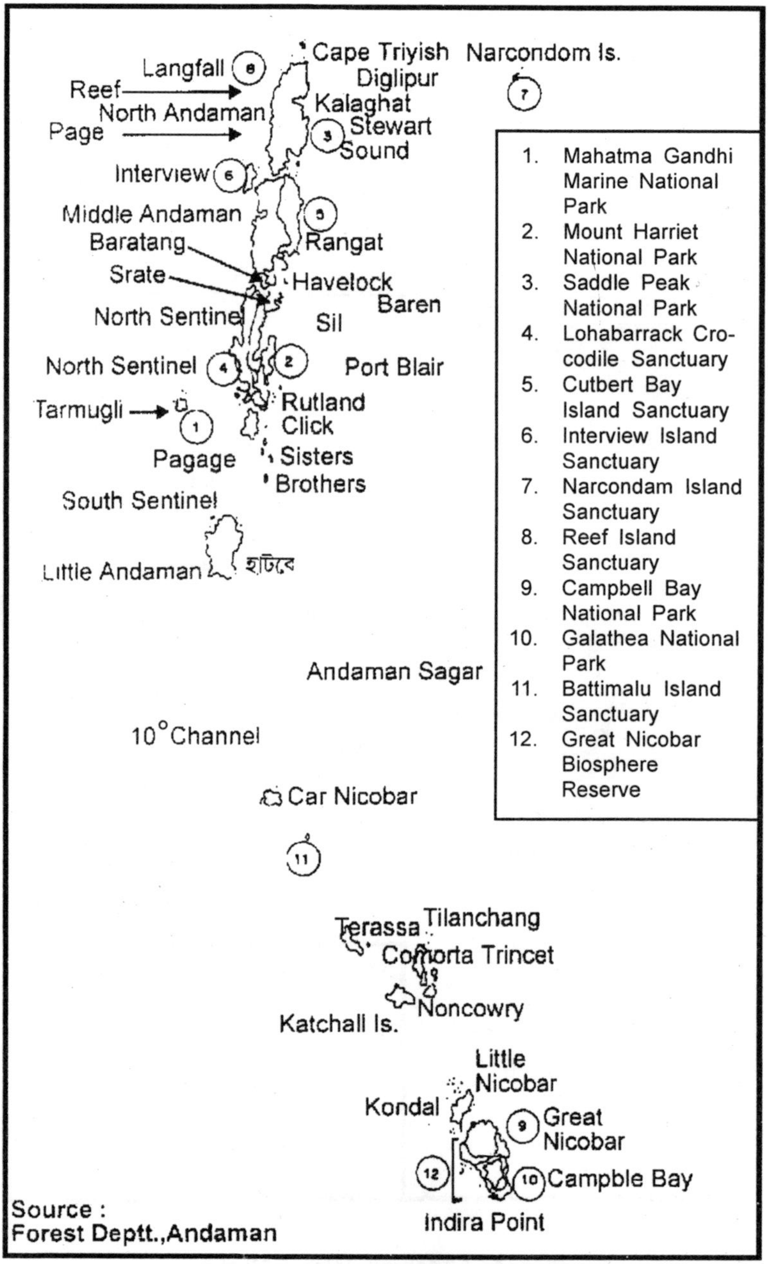

Map 3. Few Sanctuaries and National parks have been mentioned

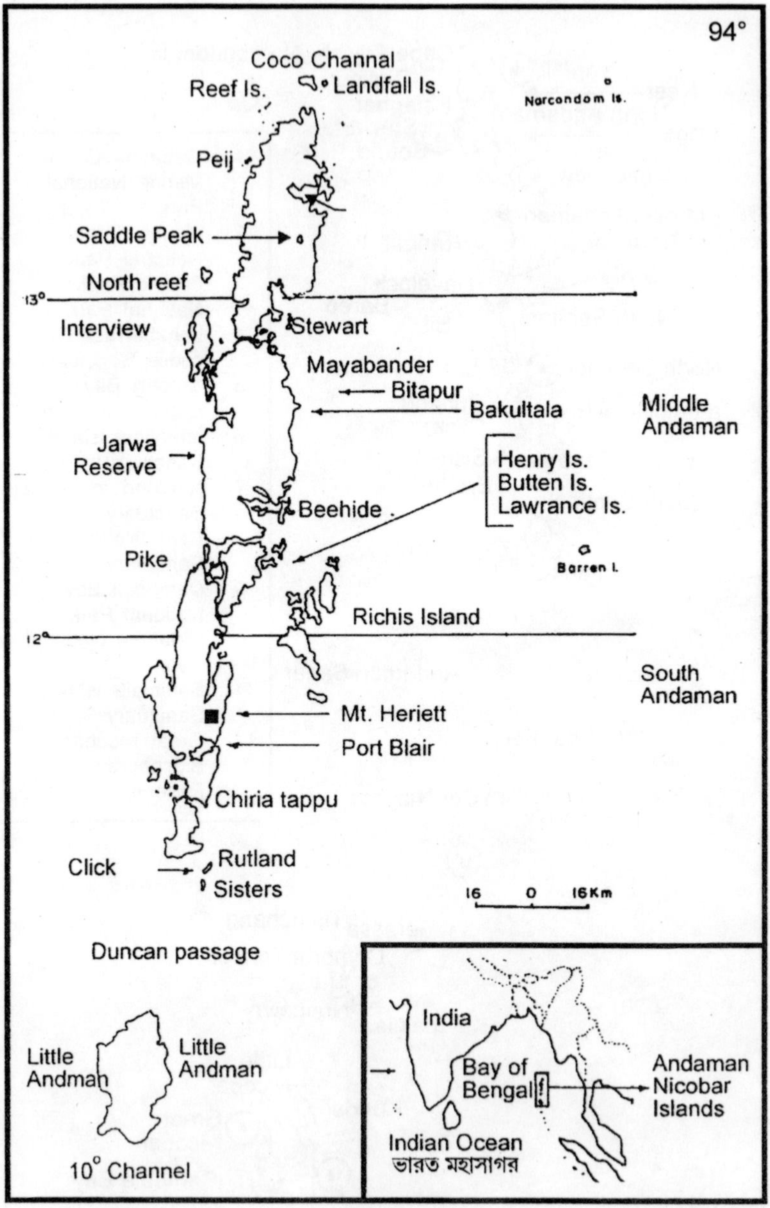

Map 4. Andaman Island Group

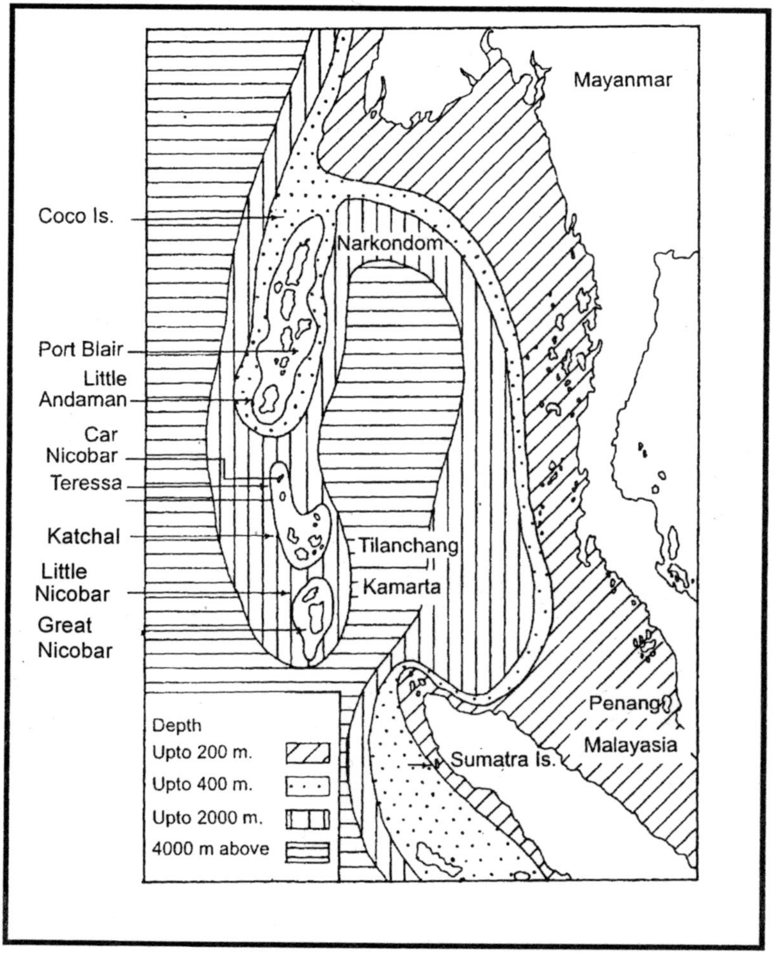

Map 5. Depth of sea continental shelf

1

Bay Islands are Learner's Living Museum

THE ISLAND'S PROFILE

Background Information

The Andaman and Nicobar Islands consist of a chain of 348 large and small islands and islets situated in Bay of Bengal. The tract lies between the latitude 13°-41° and 6° North and the longitude of 92° and 93°-7′ East. The main island runs almost in a chain parallel to North-South direction towards southeast of India. There are 324 islands in Andaman group and they extend over 754 km North-South.

There are two distinct island groups in the archipelago, the Andaman group north of ten degree channel the Nicobar group south of it. Barren island (erupted on 30.4.91) and Narcondam Island are of volcanic origin while the others are the tips of a submarine mountain extending from Sumatra (Indonesia) to Aracans in Burma.

The highest point in the archipelago is Saddle peak in North Andaman rising to a height of 732 m above sea level. The coastal strip is narrow.

The islands have a population of 1,99,254 (1981 Census). Of this, the aborigines account for less than 1000.

The total geographical area covered is 8293 sq.km of which forest areas is 7171 sq.km. Only 34 out of 348 Islands are inhabited.

Physiography, geology, soil and climate

The terrain of these islands is irregular and is more or less hilly to undulating, sometimes flat in the valleys. The hills following the general direction of the islands, lie nearly north and south. Spurs and ridges run-off the main ranges in all directions, resulting in a very confusing topography. Slopes are moderate to steep, aspect is towards east and west. The geographical configuration of Andaman Islands is described to be the Southern continuation of Arkan-Yoma tectonic unit. Geological formation of North, Middle, South and Little Andamans is mainly with intrusion of basic and ultra basic igneous rocks. The underlying rocks are essentially sandstone. Serpentine rocks are found in the higher altitude. The islands are oceanic sediments of sandstone growing into conglomerates. Andamans are a part of old Gondwana land mass, and are very ancient in origin.

The soil is mainly acidic in nature. Soil texture is medium on the surface and heavy in the deeper layer. Humus and organic matter deposition in the soil is very poor. Scarcity of sweet water is a constant problem due to the soil texture and humus content. Preservation of rain water in the rainy season is not possible and that caused erosion and due to lack of conservation of soil moisture in the dry season the soil becomes devoid of moisture.

GEOGRAPHICAL AND PHYSICAL PROFILE OF ANDAMAN AND NICOBAR ISLANDS

Total areas	8,24,900 ha. (8249 sq. km.)
Area of Andaman District	6,40,800 (6408 sq. km.)
Area of Nicobar District	1,84,100 ha. (1841 sq. km.)
Latitude	6° 45′ N to 13° 41′ N
Longitude	92° 12′ E to 93° 57′ E
Minimum temperature	16.7°C
Maximum temperature	36.1°C
Mean annual rainfall	3180 mm.

Average relative humidity 77%

Coastline 1962 km

Exclusive economic zone approx. 600,00,000 ha (6,000,000 sq. km)

Area under cultivation 15,000 ha (150 sq. km)

Area under plantation 30,000 ha (300 sq. km)

Area under forests 7,09,400 ha (7094 sq. km)

ANDAMAN AND NICOBAR ISLANDS

Special Features

Specialities of the islands are (i) isolation for thousands of years (ii) endemism of species (iii) vulnerability to outside interactions and very many other such features.

- **Peoples view the Islands as**

Floating emerald island in blue ocean – A miniature India

Abode of some rare, endangered birds – Unparallel human isolation resources

Rich continental shelf enriched in planktonic flora and fauna, fish – Abode of world's most endangered human races

Virgin rain forest which have many types, subtypes, edaphic types and associated species. – Rich museum of Coral species

- **Visitors Study Spots**

Viper island – Neil Island

Rangat – Clique Island

Mount Harriet – Chidiatappu

Sippighat – Madhuban

Ross Island – Diglipur

Havlock Island

- **Visitors attractions are:**

Coast of Karbin Cobe – Fishery Museum

Catham saw mill – Aquarium

Mini zoo – Anthropological Museum

Marine park – Cellular jail

Zoological Museum – Gandhipark

- **Other Bay Islands treasures are:**

Crab-eating macaque – Rich coral fauna

High floral/faunal endemism – Giant Robber Crab

Assemblage of several decaying tribes – Isolation for million of years

Colourful myriads of crabs mollusc and other invertebrates – Assemblage of 4 species of Sea turtles

Rare birds *Narcondam hornbill,* Megapode, Andaman teal, Nicobar Pigeon – Worlds' largest man-eating crocodile

Varied Rain Forests Varied Mangrove Forest – Vast aquatic flora and fauna

2

Biogeographical Realms

Pierre Banserean defines: "Biogeography covers origin, distribution, adaptation and association of plants and animals, their solution of changing environment and also involves recent and ancient geological history of the area." Others think that man's impact on the landscape, landuse pattern, urbanization, industry, browsing and grazing pattern, fire cycle, lumbering autecology and population structure, structure of vegetation, dynamics of vegetation, analytical phytosociology etc. cover the definition. Udvardy M.D. puts, ethology, ecology, animal ecology, systematics, synecology, autecology, pedology evolution, geology, phytogeography and even embryology and areography under the definitions of zoogeography. The qualities of the islands are unlimited and will need volumes to describe. The Andaman and Nicobar Islands are two distinct groups of islands geologically, geomorphologically, biologically and biogeo-graphically.

- The Islands are unconsolidated marine sediments composed of mud clay and sand with only sandstone, mud stone and shale as harder substratum material. There is no crystalline rock; weathering and erosion is, therefore, high; the sandy or crystalline substratum offer easy percolation and little surface run-off, whereas in muddy and clayer substratum percolation is less and run-off is high. Increase in turbidity in Continental shelf area damages planktonic seagrass flora and fauna including coral formation. There is little ground water.

- The relict of three ancient tribes viz., the Sentinelese, the Jarwas and the Onges are invaluable wealth of the islands. While the negrito tribes inhabit the Andaman groups, the Nicobar groups have inhabitants of mongoloid group.
- The assemblages of various floral and faunal elements (Indian, Burmese, Indochinese, Japanese, Chinese, Malayan, Indonesian and Indo-Pacific) circumscribed in a limited area give the island a unique status for detailed biogeographical studies.
- There is high degree of endemism among plants (more than 10%), avifauna (more than 39.2%), Insectivora (all the six species are endemic i.e. 100%), Chiroptera (46.1%), Rodenta (more than 73.3%) and several other examples may be presented. This gives the islands a distinct position comparable to Galapogus island in pacific ocean and the vulnerability of species for confinement over a limited space.
- Many animal are confined exclusively to each group of islands viz., the Andaman group and the Nicobar group, some however, occur in both the island groups.
- Carnivora is poorly represented and the poor mammalian population is an indicator of islands a distinctive character.
- The islands have five of the seven species to Testudines of the world.
- In the order Squamata (reptiles other than turtles and a species of crocodile endemism is as high as 32 per cent.
- The islands have at least 16 species of mammals, amphibians, reptiles and birds which are rare and endangered.
- The islands are rich in coral fauna, fish fauna (more than 700 species are per Sir F. Day F.R.S. puts it).
- The resources of crustacea, mollusc, prawn, lobster, edible

oyster, mussel, black-lip pearl oyster, clam, turbo and trochus, sea cucumber, sea urchin, marine turtle and other mariculture potential.

- Dipterocarps, the dominant and indicator genus of the tropical forests of Southeast Asia, occurs in Andaman group only (not in Nicobar group) while the Nicobar group has tree palms and many species common to Malaysia, Indonesia and other southeast Asian Islands. The plant association has influenced animal species and their association.
- The islands have many species of plants, animals and birds that do not occur in Indian mainland, but occur in Burma, Malaysia, Indonesia and in other South-east Asian countries.
- Owing to large extension of continental shelf (34,900 sq.km.) the islands have a wide resources of zooplanktonic and phytoplanktonic fauna and flora respectively to this region. The physical, chemical and biological character of the marine environment are therefore, changing continuously.
- The island are cyclone prone and the maintenance of vegetation cover is essential. Long periods of geographical isolation of these islands from the adjacent land masses and subsequent changes in the environment have largely contributed to the evolution of many of their endemic species and sub-species.
- The speciation in these islands has also encouraged the rich variety of isolated biotopes, favouring unchecked evolution of a variety of species.
- The distribution of higher vertebrate groups of terrestrial animals such as reptiles, birds and mammals in the islands throws valuable light of the true insular characteristics and the extent of morphogical variation due to geographical isolation.
- Coral reefs of the fringing type are available in abundance on the eastern side of the islands while barrier reefs occur on western side.

- The Andaman fauna have greater affinities with that of Indo-Chinese sub-region, while the Nicobar fauna shows more Indo-Malayan elements in the south.
- These islands have higher percentage of endemic species distributed over a small area and are always threatened from the pressure of human activities. Insular biology has special status and conservation area for a threatened main land species as "smaller the land area the higher rate of normal extinction." Mc. Arthur and Wilson (1967) in their work entitled "the Theory of Island Biogeography (Princeton University Press, N.Y.) have established the fact that the smaller the land area the higher the rate of natural extinction."

Thick Eocene sediments deposited on pre-tertiary fine grained sandstones, shales, salt stones and lime stones with instrusion of basic and ultra basic igneous rocks. Sandstones are commonest rocks, porous and slightly micaceous. In hill top there is green and red jaspers, quartzite, crystalline limestone, occasionally gneissic outcrops. Car Nicobar is formed of grey clay, sandstones and upheaved corals. In Nancowrie and Kamorta the bed rock is serpentine, covered by whitish or yellowish clay. Katchal is composed of calcarious sandstones; little Nicobar and Great Nicobar has micaceous sandstones.

In little Andaman mean annual rainfall varies from 3140 mm to 3200 mm. Mean annual temperature varies from 26°C to 27° C and maximum temperature rises to about 31° to 32° during April-May.

The Nicobar group of islands are surrounded by coral self and shallow seas of varying width, long and narrow stitch of sandy beaches. Car Nicobar and Katchal are almost flat Great and little Nicobar and generally hilly, highest elevation being 700 m.

Soil depth in valleys is only 60 cm to 90 cm, soil is porus except in clayey and limestone area. There is shortage of water in Feb. to April. Humus or organic matter is lacking.

Rainfall varies from 2750 mm to 4550 mm. There is no perennial stream except in Great Nicobar, Little Andaman and Katchal.

BIOGEOGRAPHIC LINKAGE OF THE ISLANDS

- Main Indian Linkage
- Myanmar
- Indian Peninsula
- Indochina
- Sri Lanka
- Malaysian

HIMALAYAN UPLIFT AND BIOGEOGRAPHICAL DYNAMIC

The biogeographical composition, distribution and the Himalayan uplift patterns are interlinked. Original Gondwana flora and fauna in both composition and general character were influenced and modified by the uplift of the Himalayas. It was primarily through the formation of biogeographical routes of interchange and barriers with Eurasia, and to a limited extent with parts of Africa. The Himalayan uplift established a direct connection between the ancient Peninsular Block of India and Asia in the north-east, whereas, the Peninsular block was rammed against the young Tethyan sediments of Asia forcing these sediments to flow around in the form of the Himalayan and Burmese Area.

After the disappearance of the Tethys sea the north-east and the older peninsular block came into contact with the young rock giving rise to the north Burma-Northeast Indian region. This region is the confluence of flora and fauna of the southeast Asia which migrated west and south-west India. This also helped migration of South and Southeast Chinese, Indo-Chinese and Malayan elements in India. Arunachal Pradesh, Nagaland and Manipur regions may be called floristic and faunistic gateway of India. The area holds and pivotal place in the historic process of progressive evolution of the present day flora and fauna.

The faunal and floral movement in the Himalayas forest had been from the east to the west. In peninsular region movements had been from the east and also from the west. The north-western route opened only when the entire Tethys had been completely obliterated; then migration took place from Europe, Africa and other places.

The Arunachal-Assam-Nagaland region is the western-most boundary of the range of many Indo-Chinese mammals, like certain squirrels and the eastern most limit of the distribution of many peninsular species such as the spotted deer.

The mammals of Assam, Nagaland are primarily Indo-Chinese, though there is admixture of the Peninsular and Ethiopian elements; the palaearctic montane elements also intrude in it. Several relict species of peninsular India (in western ghats) are also distributed over there.

PHYTOGEOGRAPHICAL BONANZA

(a) Andaman and Nicobar Islands have 2270 species of Angiosperms and 150 species of Pteridophytes of which 220 species are endemic. Endemism is rich in specific level (not generic). Rubiaceae has 23 species and 2 genera endemic in island. Among non endemic flora 50 per cent belong to India; Other 50 per cent are distributed to South East Asia, Burma, Thailand, Malaya and Indonesia. Of the 220 species (endemic) recorded over the year, it was difficult to locate at least 135 species; have they been extinct?

(b) About the rare and endangered plant species of Andaman and Nicobar Islands, the Botanical Survey of India has marked 123 species as rare and endangered. Some of the predominant families are Anonaceae, Palmae, Orchidaceae, Melastomaceae, Guttiferae, Euphorbiaceae, Rubiaceae. Other species belong to another 56 families. Of the Gymnosperm and Pteridophytes mention may be made of family Gnetaceae, Ophioglossaceae, Podocarpaceae and Psilotaceae.

(c) Rare and endangered non-endemic species of islands which occurs in South-east Asia number about 136. Principal families being Orchidaceae (17 species), Rubiaceae (14 species), Euphorbiaceae (11 species), Anonaceae (6 species), Cyperaceae (6 species).

(d) There are 19 Monotypic genera in these islands which show how endangered these genera are.

(e) Great Nicobar, which has no Dipterocarpus or Pterocarpus species, is very rich in various species of tropical forests. Many species have been found endemic to the island. Some endemic species are *Cyrtendra burttii, C. occidentalis, Cyrtandromoea nicoborica, Jainia nicobarica, Ophiorrhiza nicobarica, Macaranga nicobarica, Ophiorrhiza infundibularis, Phryniun paniculatum,* and *Phanera nicobarica.*

(f) In great Nicobar about 158 or about 20 per cent of species has distribution in Southeast Asia and not found in India. Some are: *Dicranopteris curranii, Microsorium plempus, Pteris griffithii, Vittaria elongata* var. *angustifolia, Syngramma alismefolia, Crepidomanes bilabiatum, Vandenboschia maxima, Lindsaea chienii* etc.

(g) In great Nicobar itself there are 33 species of orchids belonging to 26 genera. Orchid flora show close relation with Indonesia rather than Burma. In fact there are commcn genera of orchids of Thailand, Burma, Indo-China, Malayasia, Indonesia, Philippines and Sri Lanka.

(h) Many genera of Nicobar Islands such as Cyathia, Astronea, Otanthera, Cirtandra, Stauranthera, Kibara, Stemonurus and some other species do not occur in Andamans. The flora of these tropical forests are highly niche specific, speciation is being continued all along. The unique flora trapped and confined to these islands are now being subjected to large scale biotic pressure.

ZOOGEOGRAPHY (SOME SELECTED DATA)

Smith (1931, in Fauna of British India) was of opinion that the Andaman Island contain an impoverished Burmese fauna; the Nicobar Islands approximate to the Sumatran type. Reptilean fauna indicate Malayan type. Endemicity of some land and littoral animals is as follows:

	Group	No. of Species	No. of Endemics	%age
Land	Mammals	46	32	69.56
	Birds	242	95	39.25
	Reptiles	72	24	33.33
	Amphibians	18	6	33.33

(contd.)

	Group	No. of Species	No. of Endemics	%age
	Spiders	62	28	45.16
	Moths and Butterflies	366	86	23.56
	Dragonflies	36	7	19.44
	Termites	26	14	53.84
	Antlions	7	3	42.85
	Earthworms	28	8	28.57
Littoral	Reptiles	11	0	0
	Fishes	560	24	4.28
	Holothurians	68	12	17.64
	Chitons	13	1	7.69
	Hermit crab	38	2	5.26
	Copepods	172	48	27.9
	Amphipod	110	8	7.27
	Polychaetes	186	8	4.3
	Archinnelids	18	4	22.22
	Nematodes	54	6	11.11

SOME INSTANCES OF ENDEMICITY ARE:

Shrew (Order Insectivora)

Of the six species, 3 belongs to Andaman and 3 to Nicobar Islands. All of these species are endemic. The species are: Andaman island spiny shrew, Millers Andaman Spiny shrew, Nicobar spiny shrew, Jenkins Andaman spiny shrew and Nicobar tree shrew.

Bats (Order Chiroptera)

Of the reported 32 species, 12 species are endemic (6 species each of Andaman and Nicobar Islands).

The species are Narcondam shall flying fox, Car Nicobar flying fox, Nicobar lesser short-nosed fruit bat, Andaman lesser short-nosed fruit bat, Nicobar pipisterelle and several others.

Other Chiropetra of Andaman (A) and Nicobar (N) Islands groups are:

(A) Andaman Flying Fox, Malayan large flying fox, Indian flying fox, short-nosed fruit bat, Dobson's long tongued fruit bat, Blackbeared tomb bat, Malay fats vampire, Tickeli's bat, Blyth's clubfooted bat.

(N) Malayan large flying fox, Indian flying fox, Blyth's pouch bearing bat, Fulvus leafnosed bat, lesser yellow bat, bent winged bat.

Macaque (Order: Primates)

Pig tailed and Nicobar crab eating macaques are recorded. The latter is endemic and the former is of doubtful occurrence.

Carnivora (Order)

Andaman masked palm civet is endemic. The occurrence of jungle cat is doubtful.

Pigs (Order: Artiodactyla)

Andaman wild pig and Nicobar wild pigs are both endemic.

Deer (Order: Artiodactyki)

Barking deer and spotted deer are available which have been introduced.

Rats and Squirrels (Order: Rodenta)

Of the 14 species of rats, 13 species are endemic (6 in Nicobar Island group). The species are: *Rattus rattus andamanensis, Rattus rattus atidorsum, Rattus rattus flabilis, Rattus rattus holchu.*

Others in Andamans (A) and Nicobar (N) groups exclusively are:

Rattus rogersi (A), *R. stoicus* (A), *R. palmarum* (A), *Rattus pulliventer* (N), *Rattus burrus* (N), *Rattus burrulus* (N), *Rattus burrescens* (N), *Mus musculus* (A).

Marine mammals

Dugong, common dolphin and false killer are the species, of which last two are endangered.

Avifaunal resources

The Bombay Natural History Society had made a thorough investigation of avifaunal resources and Andaman and Nicobar Islands. They have identified 255 species. They were of the opinion that except for the Megapode and Nicobar pigeon, the avifaunal affinities appear to be closure to India than to Burma and Malaya, 94 species are endemic.

Endemic species (a few) are:

Andaman green heron, Nicobar Tiger bittern, Andaman grey teal, Andaman black crested baza, Car Nicobar shikra, Katchal, shikra, Andaman crested hawk eagle, Andaman pale serpent eagle, Nicobar crested serpent eagle, Great Nicobar crested serpent eagle, Andaman dark serpent eagle, North Nicobar Megapode, Nicobar blue breasted quail, Andaman blue breasted banded rail, Andaman banded crake, Andaman white breasted water hen.

Dove and Pigeon—High degree of endemism

Andaman grey fronted green pigeon, Nicobar green imperial pigeon, Andaman green imperial pigeon, Pied imperial pigeon, Andaman wood pigeon, Nicobar coockoo dove, Andaman coockoo dove, Burmese red turtle dove, emerald Andaman emerald dove, Nicobar dove, Nicobar pigeon.

Endemism in parakeet

Large Andaman parakeet, Andaman red breasted parakeet, Blyth's Nicobar parakeet, Andaman red-checked parakeet, Nicobar red-checked parakeet.

Endemism in owls

Andaman barn owl, Andaman scops owl, Andaman lesser scop owl, Nicobar lesser scops owl, Hume's brown hawk owl, Andaman brown owl, Nicobar brown hawk owl.

Endemic kingfisher

Andaman blue eared kingfisher, Andaman three-toad forest kingfisher, Andaman stork billed kingfisher, Nicobar stork billed kingfisher, Andaman ruddy kingfisher, Andaman white breasted kingfisher, Andaman white collared kingfisher, Nicobar white collared kingfisher.

Others—Andaman chestnut headed bee-eater, Andaman broad billed roller Narcondam hornbill, Andaman black woodpecker, Andaman fulvous breasted pied woodpecker, Nicobar hooded or green breasted pitta, Andaman black napped oriole, Nicobar blacknaped oriole, Andaman black headed oriole, large Andaman drongo; small Andaman drongo, Andaman racket tailed drongo, Nicobar drongo; similar 30 other species may be cited.

Endemic sunbirds

Andaman olive backed sunbird, Nicobar olive backed sunbird, Car Nicobar olive backed sunbird and Nicobar yellow backed sunbird are four species:

Winter Visitors

The following are winter visitors:

Common teal, Horse feeld's Goshawk, Asiatic Sparrow-hawk, Pole harrier, Maopagu's Harrier, Marsh harrier, Osprey, Eastern Peregrine falson, European kastrel, East Himalayan kastrel, Eastern Baillon's crane, Collared pratincole, Grey plover, Eastern golden plover, Whimbrel, Eastern whimbrel, Eastern curlew, Bartailed godwit, Common red shank, Green shank, Green sand piper, Wood sand piper, Terek sand piper, Common sand piper, Turnstone, Pintail snipe, Fantail snipe, Great snipe, Jack snipe, Eastern knot, Eastern little stint, Little stint, Temminck's stint, Longtoed stint, Curlew sand piper, Broad-billed sand piper, White winged black tern, Javan gull billed tern, Emerald Cuckoo, White cheeked Grey Drongo, Rosy pastor, Olive flycatcher, Brown flycatcher, Eastern red-breasted flycatcher, Northern bluethroat, there are also 10 species of warblers, species of Wagtail, 3 species of pipets and 2 species of Bunting.

Reptiles

Of the 80 species of reptiles and amphibia known to the islands, 31 species visit sea shore at some stage of their life (indicates necessity of protecting sea shore). Stoloczka thinks that several species of lizards and snakes (common to both the island groups) and other animals resemble Indo-Malayan, gradually passing into Burmese. Some of the species are:

Snakes (Order: Squamata)

40 species have been recorded of which brown pit viper, Andaman pit viper, Colubrine Amphibious Sea Snake, Andaman banded Krait Boie's Cat Snake, Nicobar water snake, Nicobar strippedneck snake, Oat's blind snake, Brown spotted pit viper, Cantor's pit viper, Andaman banded kukri and Andaman blink snake are endemic.

Other reptiles (Order: Squamata)

Of 33 species the following are endemic (10 species): Curltailed gecko, Andaman pay gecko, Green forest lizard, Tiwari's garden lizard, Andaman garden lizard, Andaman skink, Tytler's skink, Nicobar tree skink, White stripped skink and Andaman water monitor.

Amphibia

There are 8 species of frogs and 2 species of toads of which Pillai's Narrowmouth frog and Nicobar toad are endemic. Others are brown frog, paddy frog, Andaman paddy frog, Nicobar frog, salt water frog, Boulenger's narrow mouthed frog, tree frog, Indian toad and Nicobar toad.

Rare and endangered species

Andaman Wild pig, crabeating macaque, Dugong, Estuarine crocodile, Green sea turtle, Hawkbill turtle, Water monitor, Andaman teal, Hawks, Large falcons, Narcondam hornbills, Nicobar megapode, Nicobar pigeon, Osprey or fish eating eagle and White breasted sea eagle.

Insect fauna

Much information is not available about insect fauna of the island. Some records available in the records of Zoological Survey of India are summarised below:

Odonata	—	A.R. Lahiri (1975 Rec. ZSI); 31 species in South Andamans of which 5 are endemic, 3 common to India, 4 common to Sri Lanka and 10 to South-east Asia.
Coleoptera	—	19 species of aquatic coleoptera (aquatic beetles) have been identified.
Termites	—	Roonwal and Bose (Rec. ZSI 1964)

identified 17 Species of which 10 are endemic, 6 species are Common with oriental region.

(Genera-Neotermes, Cryptotermes, Prorhinotermes Schedorhinotermes, Hospitalitermes, Nausetermes, Odontotermes).

Spiders — BK. Tkkadar (Rec. ZSI 1977) records 41 genera and 58 species of spiders. The work involved also those of Rhorrel and Pockok.

Fish

Hora, identified (in 1938) 112 species of fish in 40 families. But Sir Francis Day, FRS estimated about 700 fish species out of which 373 species were identified. Major genera of fishes identified were Tunies, Perches, Sharks, Corangids, Sardines, Makeral etc. Tuna shoals are often found within 5° North and South of Equator. Campbel bay is 6°4′ North lat. Has this fish in between November and May. Nicobar Island has annual potentiality of Toon fish of 50,000 tonnes.

Coral

It will be an exhaustive list of mention various species of corals that occur in the area. Of the vast coral resources it may be mentioned that exploration and identification is far from complete at this stage. The following are common genera which need be mentioned:

Acropora, Montipora, Pocillopora, Staylophora, Pavona, Euphyllis, Symphyllia, Merulina, Favidae, Goniastrea, Porites, Goniopora; Fungia etc. which have more than 150 species in Andaman water.

Crustacea

Several genera of crabs are of importance which are Scylla, Uca, Sesarma, Matuta, Grapsus, Calappa, Eriphia etc. Of the hermit crab Clibanarius has several species. More than 400 species of Crustacea have been recorded in Andaman water.

Mollusc

Many species of Conus have been collected from this locality. Of the mollusc, about 800 species have so far been recorded from Andaman and Nicobar areas. However, Nautilus, Cassis, Turbo, Trochus, Tridaena, Lambis Mure, Nerita, Trocus, Telescopium, Lyncina may be mentioned of the area.

Other invertebrate animals :

Sponges	–	70 species recorded
Coelenterates	–	117 species recorded
Stony corals	–	179 species recorded
Polychaetes	–	161 species recorded
Crustaceans	–	407 species recorded
Pycnogonids	–	8 species recorded
Sipuncalid	–	20 species recorded
Echinoderms	–	331 species recorded

3

Rain Forests of the Islands

3.1 FORESTS (QUALITATIVE ASSESSMENT)

The forests are evergreen, lofty, dense, 45 m or even more high, characterized by the large number of species of trees which occur together. Gregarious dominants are rarely met with and generally two-third or more of upper canopy trees are of species individually contributing more than 1 per cent of the total number, a few species, however, do occur semi-gregariously. Some species of the top canopy trees are with clear boles 30 m. long and 5 m. or more in girth and may be briefly deciduous without affecting the evergreen nature of the forest as a whole. The canopy is extremely dense, epiphytes are numerous, especially aroids, ferns, mosses and orchids. Climbers are very conspicuous in semi-evergreen and moist deciduous forests than in evergreen. Ground vegetation is not dense. The undergrowth is often a tangle of canes, creeping bamboos and palms, plank buttress is common, the leaves are thick and glossy. Cauliflory is common. Dipterocarps are very characteristics trees in the forests of Andaman (not in Nicobar islands) and therefore these forests from western fringes of humid tropic where Dipterocarps predominate. Besides palms and canes which are so conspicuous the bamboo Dinochloa andamanica and Oxytenanthera nigrociliata occur abundantly.

The following types of forests and some principal species as classified by Champion and Seth are found in the Islands. The edaphic and climatic factors play dominant role in vegetative formations.

These plants and animals are the inhabitants of two major types of ecosystems themselves consisting of several micro-habitatas:

I. The island ecosystem, comprising of

Ia. The forests

Ib. The coasts (including mangrove, littoral, and swamp forests, beaches, and other intertidal zones)

2. The marine ecosystem, comprising of

2a. The deep, open seas

2b. The coral reefs.

At present, the A&N islands are reported to have a forest cover of 86% [Anon 1986]. According to Champion and Seth, there are 12 forest types occurring in A&N [Champion and Seth 1968] are:

- Giant evergreen forest (1A/C1)
- Andamans tropical evergreen forest (1A/C2)
- Southern hilltop tropical evergreen forest (1A/C3)
- Cane brakes (1/E1)
- Wet bamboo brakes (1/E2)
- Andamans semi-evergreen forest (2A/C1)
- Andamans moist deciduous forest (3A/C1)
- Andamans secondary moist deciduous forest (3A/C1/ZSI)
- Littoral forest (4A/LI)
- Mangrove forest (4B/T52)
- Brackish water mixed forest (4B/TS4)
- Submontane hill valley swamp forest (4C/PS2)

The Giant evergreen forest is "...the most luxuriant type of forest....In the Andamans it is met with on deep alluvial soils near the banks-of larger streams but is so intimately mixed with the semi-evergreen forest that delimitation is difficult." This forest type is generally found in the valleys. Major species included Dipterocarpus alatus.

The Andamans moist deciduous forest has a ". . . somewhat irregular top storey of predominantly deciduous trees about 40 m or more in height . . . Beneath these trees is a rather definite second storey of numerous species including some evergreens . . . and there is a fairly complete shrubby evergreen undergrowth." This forest type is more extensively distributed in the Andamans while it is not so extensively distributed in the Nicobars. The major species include *Pterocarpus dalbergioides, Terminalia bialata* and *T. procera*.

The Andamans secondary moist deciduous forest is characterized by a ". . . more or less even aged pole crop. This forest type is found in worked over areas of the primary type." The major species include *Canarium euphyllum, Pterocymbium tinctorium* and *Salmalia insignis*.

The Littoral forests occur "all round the coast wherever a fair width of sandy beach occurs, including sandy bars on the sea face of the river deltas." In the Andamans, *Manilkara littoralis* is the most characteristic species of this forest type.

The Mangrove forest is a ". . . closed evergreen forest of moderate height composed of trees specially adapted to survive on tidal mud which is permanently wet with salt water and submerged every tide." The major species indude *Rhizophora mucronata,* which occurs in the Andamans and *Bruguiera conjugata,* which occurs in the Nicobars.

The Brackish water mixed forest ". . . is the finest development of the tidal forests and may be a closed forest over 33 m. high (Bruguiera and Heritiera)." This forest type is probably not found in the Nicobars. Major species found in the Andamans are *Heritiera littoralis, Brownllowia lanceolata,* and *Phoenix paludosa*.

The Submontane hill valley swamp forest is an "irregular forest of a limited number of mainly evergreen species able to withstand the wetness of the sites occupied. The trees are usually lower owned and branchy and there is often a dense growth of Calamus, while Ficus and monocotyledons such as Alpinia often form the undergrowth."

The forests of A&N are significant not only because quite a few plant species found in them are endemic, but also because a few others are not found on the Indian mainland, but are closer to Burmese and Malaysian flora [Thothathri 1962]. This may be due to the geographical proximity of the islands to Burma and Malaysia. Among the species occurring here whose main distribution is Burma, are: *Anaxagorea luzoniensis, Dipterocarpus incanus, Ancistroctidus tectorius*, and *Olax imbricata*. Those which are commonly found in Malaysia, and occur in A&N are *Pyrrosia longifolia, Scolopia spinosa, Ficus gibbosa, Ficus rumphii*, and *Clerodendrum viscosum*.

There are interesting differences in vegetational composition between various islands of the Andamans, between various islands of the Nicobars, and between the Andamans and Nicobars [Thothathri 1962].

South Andaman forests have a profuse growth of epiphytic vegetation mostly ferns and orchids; the Middle Andamans harbour mostly moist deciduous forests; North Andaman is characterized by wet evergreen type, with plenty of woody climbers. The northern islands of the Nicobars (including Car Nicobar and Battimalv) are marked by a complete absence of evergreen forests, while such forests form the dominant vegetation in the central and southern islands of the Nicobars. Grasslands occur only in the Nicobars, and while deciduous forests are common in the Andamans, they are almost absent in the Nicobars. Several species are endemic to either of the two island groups-the Tree fern *Alsophila albosetacea* to the Nicobars, the major timber species Dipterocarpus spp. (Gurjan) and Pterocarpus dalbergioides (Padauk) to the Andamans.

The author was associated in a work of Government of India which has published 'Status Report of Mangroves' (Anonymous, 1987) and stated that in the Andaman Group more than 350 islands and the total area of all these islands are estimated about 8,500 km^2. While in Nicobar group, the total lands are estimated about 1,654 km^2 within 30 islands. The 1986 km long and deeply indented coastline of these Andaman & Nicobar Group of Islands provide innumerable creeks, bays, estuaries and these facilitate the development of rich and extensive mangrove forests. This forest area constitutes 9.4 per cent of the total land area and about 10.9 per cent of the total forest area of the Andaman & Nicobar Group of Islands (Dagar, *et al.*, 1991).

These mangrove zones in the Andaman & Nicobar Islands are estimated about 18 per cent of the Indian total mangals (Naskar & Guha Bakshi, 1987 and Anonymous, 1992). Jagtap (1985) has reported only 20 mangroves and two mangrove associates from these Andaman Group of Islands, viz., *Rhizophora mucronata, R. apiculata, Bruguiera gymnorrhiza, B. parviflora, B. sexangula, Kandelia candel* (Syn. *K. rheedii*), *Ceriops tagal, Aviceras corniculatum, Acanthus ilicifolius,* Derris sp., *Excoecaria agallocha, Heritiera littoralis, Nypa fruticans, Acrostichum aureum, Scyphiphora hydrophyllacea, Clerodendrum inerme, Thespesia populnea.* Jagtap, *et al.* (1993) have further reported 27 mangrove flora from Andaman & Nicobar Islands. In another publication Jagtap (1992) has studied and collected mangroves from Car Nicobar, Katchal and Nancowry and also reported mangroves from this Nicobar Group of Islands, viz. *Rhizophora mucronata, R. stylosa, Nypa fruticans, Sonneratia alba, Lumnitzera littorea, Heritiera littoralis, E. agallocha, Derris heterophylla, Bruguiera gymnorrhiza, Acrostichum aureum.*

These mangrove areas of the Andaman & Nicobar Group of Islands are estimated by different workers, differently as follows:

Mathuada (1957) - 1,15,200 ha (28,800 acres); Waheed Khan (1957) - 1,20,900 ha (1,209 km^2); Sahni (1957) - 1,17,100 ha. Qureshi (1957) has estimated, the mangrove area 1,152 km^2 in the Andaman Group of Islands and 35.33 km^2 in the Nicobar Group of Islands. Sidhu (1963), Blasco (1975), Untawale (1987), Government of India-Mangrove Status Report (1987), Balakrishnan (1989), Working Plan of A & N Forest Department, Dagar, *et al.* (1991), Jagtap, *et al.* (1993) have also separately reported mangroves of Andaman 1,15,200 ha., 1,15,000 ha., 1,15,000 ha, 1,190 km^2, 1,01,172 ha, 77,769 ha, 82,251 ha, 780 km^2, respectively. Recently (1992), the Space Application Centre (ISRO), Ahmedabad has reported the mangrove areas in the Andaman & Nicobar Islands as 770.5 km^2 in the following regions, viz.

(a) **North Andaman:** South of landfall island, creeks on the Bheele bay, the Elizabeth bay, the Duncan bay, the Katara creek, the Reef inlet, the Hudson bay, the Gudinala, the

Coffrie bay, the Mc Pherson bay, the Mangrove bay, the Blair bay, the Gibb creek, the Bacon bay, the Shallow bay, the Godwin bay, the Luis inlet, the Robert bay;

(b) **Middle Andaman:** on the Qutram Island, the Herry Lawrence Island, along the Rangat bay, the Austen strait, the Cook's passage;

(c) **South Andaman:** the Corieve bay, the lagoon in port Meadows, on the Wilson island, along the Blair reef, the Constance bay, the Shoal bay creek, the Flat bay, along the creek on the Havelock Island;

(d) **Little Andaman:** along the Bumila creek, the Egubelong creek, the Jacson creek, Dugond creek and the Tae-eye-creek;

(e) **Nicobar Group of Islands:** near Cape Maud on the Tilanchang dwip, along the Satellite bay on the Comorta Island, along the creeks on the Trinkat Island, the Nancowry Island, along the East and West Bay on the Katchau Island, Car Nicobar and Morgan Nala of the Cambel bay (Great Nicobar).

According to the estimate and reports of Dagar, *et al.* (1991) the mangrove flora are 34 in the Andaman & Nicobar Group of Islands; these are as follows: *Bruguiera cylindrical, B. gymnorrhiza, B. parviflora, B. sexangula, Ceriops tagal, C. decandra, Kandelia candel, Rhizophora apiculata, R. mucronata, R. lamarkii, R. stylosa, Avicennia marina, A. officinalis, Sonneratia alba, S. apetala, S. caseolaris, S. griffithii, Heritiera littoralis, Xylocarpus granatum, X. mekongensis, X. moluccensis, Aegiceras corniculatum, Aegialitis rotundifolia, Scyphiphora hydrophyllacea, Lumnitzera littorea, L. racemosa, Excoecaria agallocha, Cynometra iripa, C. ramiflora, Nypa fruticans, Phoenix paludosa, Acanthus ilicifolius, A. ebracteatus, A. volubilis*.

These mangrove flora and the mangal zones of the Andaman and Nicobar Islands are least disturbed and best preserved in the perspectives to the mangals of other Indian territories.

Since the year 1870, Sultiz Kurz, Kloss (1920), Parkinson (1923), Suri (1928). Curtis (1933), Chengapa (1944), Sahni (1953, 1958 & 1959), Banerjee (1959), Thothathri (1960 & 1962), Blasco (1975), Balakrishnan (1979), Untawale (1984), Rajagopalan (1987),

Jagtap (1981 & 1992), Dagar *et al.* (1982), Jain (1980) have studied the different floristic aspects on the Andaman & Nicobar Islands. Dagar, *et al.* (1991) have published the research data on the mangroves of the Andaman & Nicobar Islands very critically.

[*Source*: aforesaid few paragraphs are from Naskar & Manldal's "Mangrove Ecosystem"]

ECOLOGICAL STATUS OF FORESTS

There are various views. Some place it as 'Climatic Climax' some as 'Edaphic Climax' and some as 'Seral'.

The authors (1972) classified the formation of vegetation in the West Bengal Sundarbans as detailed in the statement No. 1. Another classification has been made on tide level also keeping the formation unchanged which may be seen in another statement No. 2. The statement also gives the density of crop. A series of profile diagrams have been given to show the two vertical serata of forests in each formation.

Suaeda maritime and S. nudiflora, Sesuvium portulacastrum, Acanthus illicifolius and Acrostichum aureum form saline vegetation in partly reclaimed areas where sweet water species invade. Invasion of species has been recorded in detail and enumerated in the family, genera and species list.

While Heritiera spp. and Excoecaria spp. dominate in Bangladesh, Excoecaria (dwarf form) and Ceriops sp. dominate in Indian Sundarbans. Avicennia is rather poorly represented. It is probably a 'Seral' species.

There is no humus on Sundarbans forests floor. The flat lands that grow higher and higher by silt deposition sometimes create large flattish and shallow depressions known as maidan or blanks (Saline blanks) where owing to absence of inundation solar and aerial evaporation leads to concentration of salt. The status of such land is yet to be determined.

Stamp (1925) while studying Burma mangroves identified the forests as 'Climatic climax community'. Tansley and Chipp (1926) considered mangrove forests as an 'Edaphic community'. Champion (1936) placed such forests under 'Primary Seral type' of moist Tropical Seral formation. But Davy (1938) placed mangrove under Woodland formation of tropics which is climatic climax.

It is necessary to study such forests from aerial photographs followed by ground survey. From aerial photographs the following has been ascertained on the distribution of Mangrove forests in Andaman and Nicobar Islands.

North Andaman	29254 ha.
Middle Andaman	23116 ha.
South Andaman	21389 ha.
Other areas of Andaman & Nicobar Islands	41441 ha.
Total	1,15,200 ha.

Middle and South				
Andaman	2448.71	87.7	371.12	12.3
North Andaman	1191.50	72.8	272.20	16.6
Nicobar Islands	592.62	91.5	12.89	2.0
Total	4886.81	87.6%	693.80	12.4%

According to State Working Plans Mangrove Forest area has been identified as 1,15,200 ha. which also involves ecotone areas. There might be quite a lot of shrinkage owing to felling of trees for firewood.

The statements that follow in Tables I to IV give the summary of findings.

Table I: Andaman and Nicobar Islands Land-use pattern in North Andaman from Aerial photo interpretation

Particulars	GENERAL Area	Percentage
(a) Total area interpreted	1636.91	100
(b) Forest Area	1517.58	92.7
Plantations	20.17	1.2
Scrub	25.90	1.6
Scrub and bamboo	7.79	0.5

Particulars	Area	Percentage
	GENERAL	
(c) Non-Forested area	119.55	7.3
Forest Types		
(a) Evergreen, Semievergreen and Deciduous forests	1191.50	72.8
(b) Mangrove		
Forests...Rhizophora species	118.64	
Bruguiera sp.	83.83	
Mixed Crop	69.73	16.6

Table II: Land-use pattern in Nicobar Islands from aerial photo Interpretation

Particulars	Area	Percentage
	GENERAL	
(a) Total area interpreted	647.15	100
(b) Forest area	611.92	94.6
Plantations	19.30	3.0
Scrubs	0.60	0.1
(c) Non-Forests	25.23	5.4
Forest Types		
(a) Evergreen, deciduous and Semi evergreen forests	592.62	91.5
(b) Mangrove Forests leaders		
Rhizophora	3.50	0.6
Bruguiera	3.96	0.6
Mixed	5.43	0.8

ISLANDS UNIQUE ECOSYSTEM

Ecologically tropical islands have certain unique assets but they are also extremely vulnerable to manipulations. Their location in the tropical zone bestows upon them climatic and biotic conditions best suited for natural high productivity. Uniform warmth round the year, uniform day light length and plentiful precipitation maintaining high humidity results in luxurious vegetation growth. In fact the volcanic tropical islands like Java have the world's highest density of population maintained by the richness of the productivity of land.

Table III: Land-use pattern in Middle and South Andaman

GENERAL		
Particulars	Area	Percentage
(a) Total area interpreted	3026.68	100
(b) Forest area	2819.83	93.5
Plantation	93.97	3.1
Scrub	16.61	0.6
Scrub and Bamboo	8.40	0.3
(c) Non-Forested area	206.85	6.8
Forest Types		
(a) Evergreen, Semievergreen and Deciduous forests	2448.71	87.7
(b) Mangrove forests leaders		
Rhizophora	201.31	6.6
Bruguiera	31.91	1.1
Mixed	137.90	4.6

Table IV: Land-use pattern in Little Andamans as interpreted from Aerial photo interpretation.

GENERAL		
Particulars	Area	Percentage
(a) Total area interpreted	679.82	100
(b) Forest area	656.47	96.6
(c) Plantations	0.05	-
(d) Scrub	2.44	0.4

Wherever human interference has been minimum, the extremely productive coral reefs and mangrove ecosystems on the coastal seas and the complex tropical rain forests on dry land have remained intact. Smaller tropical islands never had settled agriculture as a traditional pattern of existence for the indigenous populations. Collecting marine produce or hunting-gathering were the life styles that evolved in tropical islands. Cultivation of land even if it occurred in the larger islands depended on shifting cultivation where nutrients are recycled by partly burning the primary vegetation and after short-term cultivation permitting natural regeneration to rebuild soil. An alternative land use is flood plan cultivation where annually floods deposit humus rich silt brought down from the river catchment hill forests. The traditional fishing-hunting-food gathering, communities needed a large extent of habitat to obtain their requirements without depleting the resource. The viability of slash and burn cultivation similarly demanded low population densities (2.3 people sq.km) and long rest periods after 2-3 years cultivation. Only in exceptional cases where constant volcanic activity provides inexhaustible mineral nutrients, intensive cultivation has been possible as in Java or other parts of East Indies islands.

The very same tropical ecological conditions favourable for biotic productivity also imposes a degree of vulnerability to these islands. Heavy rainfall often results in very impoverished soils because of the continuous leaching action of water. Higher ambient temperature and intense bacterial activity results in rapid decay and release of nutrients locked up in plant and animal body. The soils are shallow, poorly formed and extremely fragile depending on the climax vegetation clothing it for protection. The islands being smaller in area will not have extensive drainage basins where erosion-deposition cycles can create deep soil layers. Tropical rains are of such intensity (more than 40 times that of temperate regions) that is erosive power is tremendous. As much as 185 tons of soil have been reported to be washed out from one hectare of denuded land in one storm. Larger rain drops without vegetation cover shielding the soil result in soil compaction reducing percolation which in turn results in greater surface erosion.

Normally, the multi-layer forest canopy over tropical soils breaks up three fourth of the rainfall into a fine spray and the remaining one-fourth tickles down the tree stem enriched by nutrients in the bark to reach the soil. The decomposition rate in the humus-litter layer is precisely balanced depending on the insulation of vegetation cover that keeps the soil temperature around 25°C. But once the canopy is opened up, temperature goes above 27°C, and the decay loss is more than production. Destruction or serious disturbance to forests immediately results in impoverishment of soil. In fact, in tropical forests, the root system is exceptionally, well developed forming a mat immediately below the litter and actually above the soil. Deeper soil horizons will have practically no roots.

None of the cultivation patterns of the 'civilized' human societies are suited for the tropical rain forest soils. No monoculture, particularly after removal of canopy forest can remain viable in the long run. Cash crop plantations and food crops equally get affected by pests and pathogens quickly in the unnaturally simplified agro-systems of man. Tropical soils support an extremely species-rich complex rain forest which in turn produce and protect the soils. The most essential attribute of the tropical forests is its complexity, containing thousands of species of plants per hectare but at the same time each species has only a low population density (often only 2-3 individuals per hectare). This diversity assures optimal use of all environmental conditions, but it entails that only large unmodified areas can assure species survival. Yet another characteristic, having a crucial role in the survival of tropical rain forest ecosystems is the extremely complex interdependence of different species in this biotic association. Plants and animals are so intricately linked together that it is next to impossible to identify the various mutual inter-dependencies. The role of same microscopic mycorrhiza for seed germination of a tree or the essential role of a species of wasp in pollinating a plant are easily overlooked but they are extremely critical for concerned species and indirectly for the whole interdependent system.

Any modification of this intricate system by man through forestry, over-exploitation of specific species, clearing for cultivation etc. undermines its stability. In spite of its tremendous potential richness, it remains practically unknown to Science. Of the potential 5 million species of plants and animals the planet may have, only around 1.5 million have been recorded by science and of this less than 10% have been screened for potential use by man. Yet conservative estimates consider that by A.D. 2000, 1 million species would become extinct mostly through destruction of tropical forest areas.

The shallow seas surrounding tropical islands harbouring extensive coral reefs and the interface of the sea and land, in suitable areas covered by mangrove forests are also reflections of the potential productivity of tropical ecosystems unmodified by man.

The organic and mineral components in the soil of tropical forests ultimately get washed down to the sea and may be lost for ever in the sea bottom sediments but for the coral reefs and mangroves. The very complex marine animal community that we call the coral reefs are basically built up by very small invertebrates—the coral building coelenterates. These specialized organisms need unvarying salinity, temperature and light conditions so that they can establish themselves only near the equator in shallow seas with no turbidity. Using these coral foundations a manifold animal society comes to get established and perpetuates itself. This ecosystem is an exact counterpart of the tropical ever-green forest in productivity, complexity and diversity. But the coral reefs are very sensitive ecosystems. They are extremely finely balanced. Increased turbidity in the coastal waters due to accelerated erosion by inland streams, as a consequence of deforestation, for example, can destroy coral reefs in no time. Only live reefs keeping pace with tide and wave damage can maintain itself. If fresh coral growth is arrested, reefs disintegrate and disappear in no time. Naturally this in turn will affect the nearly coast also in addition to the biota it supports directly.

Mangroves similarly are a specialized plant association supporting a wealth of animal life that has evolved in tropical coastal areas where rich alluvial deposits accumulate. Due to a variety of physico-chemical-environmental conditions such as fluctuating salinity, permanently water-logged soil, soil acidity etc., only mangrove species can thrive in such areas and maintain the high level of biotic productivity, the high tropical solar energy permits. But just like the coral reefs, the mangroves are also sensitive ecosystems. Any modification of the drainage pattern altering the hydrological cycle, in turn affecting soil salinity and water content, will adversely affect the mangroves. Because of plentiful availability of food and shelter otherwise lacking in the featureless coast, the mangroves attract a host of marine as well as brackish water life forms. Many of these use the mangroves at least during part of their life cycle. Destruction of mangroves hence not only eliminates obligatory residents of this ecosystem but also seriously disrupts littoral fish populations as well as inland fresh water fishes whose larvae at least use the mangroves.

The smaller the size of the Island, the more closely inter-linked the components of the island ecosystems such as dry land forests, coastal mangroves and coral reefs. Disturbance in any of these will immediately affect the whole complex.

The Andaman and Nicobar islands at least till recently were entirely covered by dense rain forests, supporting only a small pre-agricultural aboriginal population. Here we have an opportunity to understand the richness and fragility of valuable ecosystems and protect them for mankind. All the earlier mentioned ecological constraints to development of agriculture or forestry apply here. Added to these is the geological peculiarities. The entire group of islands, excepting Narcondam and Barren islands are uplifted marine bottom sediments with some secondary calcareous formations or a few ultra basic intrusions of hard rock. The clayey or sandy or porous calcareous substratum reduces nutrient availability as they have little mineral content needed for plant growth. The porosity or total impervious nature rules out ground water storage and controlled run off. Erosion is unusually severe on this soft basic material. The steep slopes of the islands aiding rapid run off along with soil unsuited to retain rain water result in non-availability of fresh water during the non-rainy period. The natural eco-systems which have colonized these peculiar environments are very specialized so that the mangrove or littoral forest cannot be replaced by any other man-made system.

In comparison with all the other tropical moist forest of the world namely the Amazon basin and the Central African belt, the S.E. Asian belt has some of the oldest and hence stablest and maximally diverse natural forest areas. These Pleistocene refugia of plants and animals have an unbroken history of at least 60 million years of *in situ* evolution. Yet they are shrinking rapidly. About 10 million hectares of tropical humid forests are cleared each year (at the rate of 40 hectares per minute) and in S.E. Asia, 1.8 million hectares are disappearing per year. In India, these forests restricted to the North Eastern region, the Western Ghats and the Andaman and Nicobar islands are in a critical phase. They are practically totally fragmented or modified in the Western Ghats, excepting small pockets like the Great Nicobar Island, the Little Andaman and some smaller islands, viable units no longer occur. This is all the more reason to protect the remaining areas of the island ecosystems. In other parts of India a total moratorium on all tree felling may not be feasible however desirable it be, due to the very large population continuing to be directly dependent on forests for sustenance and where during early years of development, heavy investment has been made on forest resources based industries, which may not be able to find alternate resource in too short a time. But in the Andaman and Nicobar islands where these sociological and economic compulsions are minimum, it is possible to save the remaining forests from total destruction.

[*Source*: From Dr. S. Nair's report.]

4

Physiography of Coastal Bonanza

Thousands of islands, creeks, inlets provide kaleidoscopic view of forests, scrubs, varied coastal types, tidal actions, dancing of crabs and migration of molluscs. Tourists will find a wide opportunity to enjoy these landscape and study the behavioural pattern of these animals. A special point which features in the area is that various facets of sea coast can be viewed and studied without involving a long journey or making any pre-arrangement and involvement of heavy men and material for such study.

DIVERSE SEA COAST

(Visitor's wide study zone)

– Fringe Mangrove (1,00,000 ha)	– Marshy reef
– Clayey store	– Sandy shore
– Marshy mangrove	– Dead coral mixed with sand
– Stoney shore	– Slashy mangrove
– Admixture stone and sand	– Stoney reef
– Numerous creeks	– Tidal Zones Subtidal Intertidal Supratidal

Sandy Coast

- Orchestea sp.
- Also occur: Fiddler crab.

  Cardius sp.

  Clam (Nectra sp.)

  Cultelus sp.
- Less species on unsettled sand
- Robber crab (Birgus latro)

- Tellina sp., Corystes sp.
- Abundance of Hermit crab (Coenobita sp.)
- Ocypode sp. (Ghost Crab)
- More species in sheltered creek

[*Source*: Das and Deb Roy, ZSI-1989 for Species lists]

Stoney Coast

- Littorina sp.
- Nerita sp.
- Barnacle Limpet

- Abundance of Algae on stone surface
- Abundance of invertebrate sp.

[*Source*: Das & Deb Roy, ZSI-1989 for Species list]

Clayey Coast

- Mangrove shelters mollusk/crab
- Sulphur bacteria and Sulphurated hydrogen
- Cerithedia crab
- Littorina undulata
- Nerita crab

- Ocean current is less
- Abundance of Fiddler crab
- Abundance of periwinkle (*Littorina scabra*)
- Abundance of invertebrates
- Barnacle mollusc

[*Source*: Das & Deb Roy, ZSI-1989 for Species list]

Tide and biota in the sea coast

This area always Remains under water	This area remains under water only twice a day	This area does not remain under water

Conspicuous plants of sea coast

- *Gnetum ula* (rare plant)
- Stenoclena palustris (a fern)

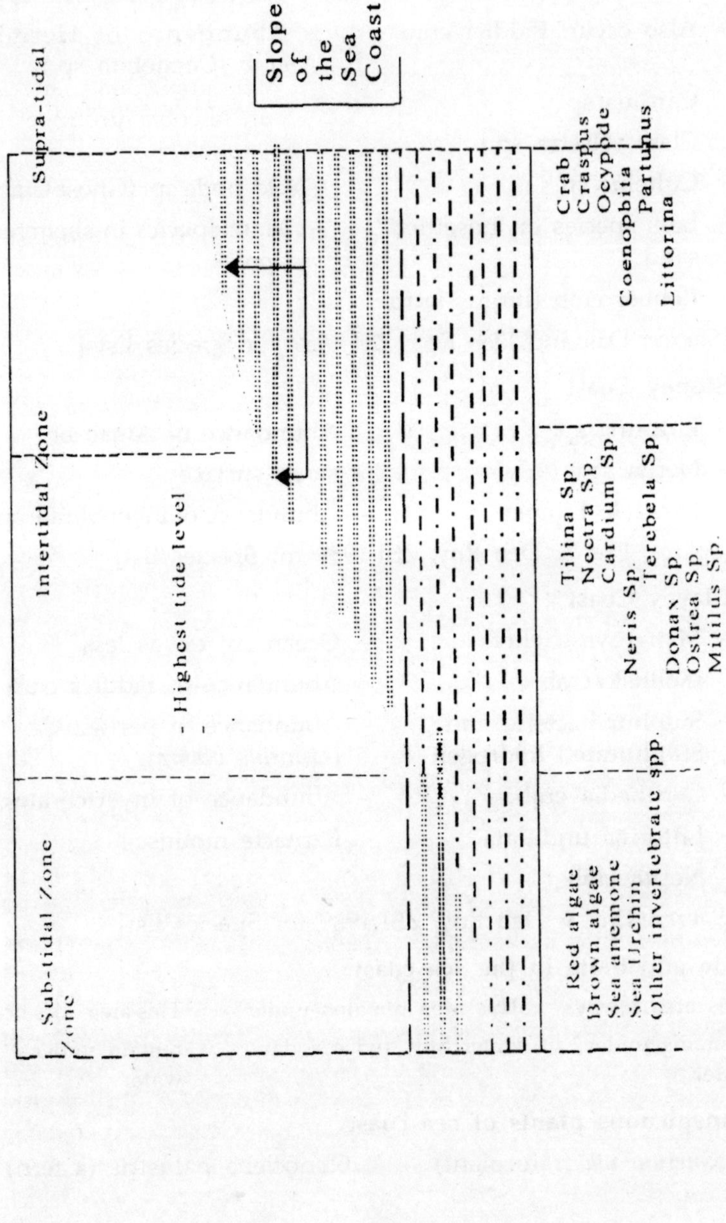
Supra-tidal
Intertidal Zone
Sub-tidal Zone
Highest tide level
Slope of the Sea Coast
Crab
Craspus
Ocypode
Coenopbita
Partunus
Littorina
Tilina Sp.
Nectra Sp.
Cardium Sp.
Neris Sp.
Terebela Sp.
Donax Sp.
Ostrea Sp.
Mitilus Sp.
Red algae
Brown algae
Sea anemone
Sea Urchin
Other invertebrate spp.

– *Cycas rumphii* (rare Plant)	– *Mjytrifera andamanica* (yellow flower)
– *Strobilanthes glenevules* (yellow flower)	– *Gymnea tingens* (yellow flower)
– *Glycosmis pilosa* (White flower)	– *Calamus andamanicus* (a climbing cane)
– *Streblus asper* (edible flower & Fruit)	– *Ochna wallichii* (yellow flower)
– *Ficus gibbosa* (edible leaf and fruit)	

Coastal gigantic tree	**Gigantic trees (Close to coast)**
– *Manilkara littoralis*	– Didu, white dhup, Garjan, Padauk, *Manilkara littoralis* Papita and Badam
– *Casuarina equisetifolia*	
– *Pongamia pinnate* (mauve flower)	

LITTORAL FORESTS

(Conspicuous species)

– *Bombax insignis*	– *Casuarina equsetifolia*
– *Afzelia bijuga*	– *Manikara littoralis*
– *Terminalia catapa*	– *Barringtonia asiatica*
– *T. bialata*	– *Erythrina veriegata*
– *Pongamia pinnata*	– *Calophyllum innophyllum*
– *Thespesia populnea*	– *Cycas rumphii*
– *Hibiscus tiliaceous*	– *Pandanus tectorius*
– *Phoneix paludosa*	– *Phoenix paludosa*
	– *Ipomoea pes-caprae*

Conspicuous herbs of sea coast

Mighty trees	Mighty trees of adjoining island
– *Manikara littoralis*	– *Salmalia insignis* (Didu)
– *Elaeocarpus robustus*	– *Canarium euphyllum*

– *Casuarina equisetifolia*

Trees of medium height

– *Pongamia pinnata*

– *Ficus gibsoa*

– *Streblus asper*

(White dhup)

– *Dipterocarpus* sp. (Garjan)

Pterocarpus dalbergioides (Padauk)

– *Pterocymbium tinctorius* (Papita)

Terminalia procera (Badam)

Composition of typical Mangroves (27 sp. Over 2000 sq.km.)

(Coastal vicinity)

– *Candelia* sp. (1)

– *Acrosticum* spp. (2)

– *Acanthus* spp. (3)

– *Lumnitgera* spp. (2)

– *Aegiceros corniculata*

– *Nipa fruticans*

– *Phoenix paludosa*

– *Bruguiera* spp. (4)

– *Ceripos* sp. (1)

– *Sonneretia* (3)

– *Heritiera* sp. (1)

– *Excoecaria* sp. (1)

– *Rhizophora* spp. (3)

– *Avicennia* spp. (3)

Plants of Andaman Mangrove

(Coastal vicinity)

Acanthus ebracteatus

A. licifolius

A. volubilis

Cerbera manghus

C. adollan

C. floribunda

Borownlowia lanceolata

R. stylosa

Aegiceros corniculatum

Nypa fruticans

Phoenix paludosas

Bruguiera gymnorhiza

B. cyhindrica

B. parviflora

B. sexangula

Ceriops tegal

Rhizophora apiculata

R. mucronata

Kandelia kendel

L. racemosa

Excoecaria agallocha

Acrosticum aureum

Xylocarpus granatum	*X. moluccensis*
A. speciosum	*Avicennia valba*
A. marina	*Sonneretia apetala*
A. officinalis	*S. alba*
S. caseolaris	*Heritiera littoralis*

5

Coastal Animals

A profile of most of the important macro animal groups have been presented in charts. The ecoeducation tourists will get enough information to pursue their studies on various species they are interested.

NATURAL ECOSYSTEMS AND WILDLIFE

With a coastline of almost 2000 km., an Exclusive Economic Zone (EEZ) consisting of 6 lakh sq.km. of marine waters and some of India's richest rainforests, the A&N Islands are the repository of an immense variety of ecosystems and wildlife. In spite of the fact that the biological investigations of the islands are far from complete, the tally of species and subspecies belonging to each class/group of plants and animals is impressive over 2200 flowering plants and 120 ferns (besides other plants), 58 mammals, 242 birds, 83 reptiles, 10 amphibians, 750 fishes, 326 coelenterates (Corals, sponges sea anemones, jellyfish, etc.) 407 crustaceans (crabs, lobsters, shrimps, barnacles, etc.), 941 molluscs (limpets, oysters, clams, snails, slugs, squids, mussels etc.), 1550 insects, 62 spiders and scorpions, and others (Rao 1989; Anon 1986). What is also significant is that a high percentage of these species and subspecies are endemic, i.e. not found anywhere else in the world. This is largely because of the isolated nature of the islands. Of the flowering plants, for instance, around 200 (almost 10) are endemic. Interestingly, another 1300 are not found on the Indian mainland, but occur in Malaysia, Burma, Indonesia, and the Polynesian islands [Anon 1986].

Animals of coastal mangrove

Bats	**Wild Pigs**
– *Pteropus melanoptus*	– *Sus scrofa andamanica*
– *Cinopterus gigateus*	– *Sus scrofa nicobarica*
– *C. phinsise*	
– *Macaca fascicularis* (Crab-eating monkey)	– *Paguma larvata* (Masked civet)
– *Tapaia nicobaria* (Tree shrew)	

Mangrove adjacent Animals of Andaman areas

Mammals (55)

– Macaque (2), Tree Shrew (2)	– Ants
Terrestrial shrew (4) Jungle	
Cat (1) Pigs (2), Plam	– Mites
Civet (1 Deer (2) Bats (26)	
– Rats (14), Squirrel (1)	– Centepedes (17)
	Spider (62)
– Birds	– Millepedes
– Lizards	– Scorpions
– Amphibia (10)	– Termites
– Snakes (46)	– Sponges (70)
– Coelenterates (147)	– Earthwoom (21)
– Crocodiles (1)	– Echinoderms
– Polychetes (161)	– Fish (750)
– Crustacea (407)	– Marine animals (3)
– Mollusca (941)	– Leeches (2)
– Insects (1500)	– Corals (179)

Diverse flora and fauna and tribe

• Excellent growth and size of Padauk and garjan Ongi and Sentinelese etc alongwith innumerable associates	Rich Corai ecosystem both the groups have their near extinct human races, e.g. differences in linkage Jarwa, High endemism

	Rare and endangered species Cyclone prone islands
• Excellent varieties of mixed miscellaneous plant species in Nicobar group of islands	Luxurious growth of edible snails lobsters, prawn, clam, Tabeir and Mussel etc.

Molluscs of sea coast

There is a rich assemblage of species. These are dominant both qualitatively and quantitatively. There are 66 species of gastropods and 32 species of Bivalves.

They inhabit the tree canopy, dead tree stumps, rocks, stones, brackish water pool and muddy substratum.

The species varieties are exhibited in 100 terrestrial species, 51 fresh water species and 90 marine species.

Various substrata	**Species of Molluscs**
Saline	*Teredo furcifera, Nausitora, Bankia, Mortesia, Lyrodas* etc.
Mud Surface	*Nassarius Cerithium Stenothya Assiminea, Neritina* etc.
Dead stumps & roots	*Onchidium, Naquelia, Nerita, Saccostrea* etc.
On living log	*Littorina scabra, Cerithidea Quadrata.*

Important general of Gastropoda (Mollusca)

Neritina	Turbo
Nerita	Littorina
Assiminea	Cerithidae
Thiara	Cerithium

Important genera of Bivalve (Bivalvia)

Nausitora	Teredo
Avacera	Perna

Donax	Crassostrea
Bankia	Gilonia
Balissa	

Crustacea of sea coast

Crustacea

Represented by Prawns, Amphipods, Isopods, Stomatopods, Hermit Crab, Procellanid Crab and Branchyuran Crab.

– Prawn and Shrimp genera	– Palaemon, Penaeus, Metapenaeopsis, Caridina and Acetes.
– Isopods and Amphipods	– Sphaeroma and Cirolana
– Stomatopods	– Gonodactylus
– Procellanid Crab	– Petrolisthes
– Brachyuran Crab	– 31 spp. and 18 genera e.g., Uca, Ocypode, Dotilla, Macrothalmus, Grapsus, Metopograpsus, Sesarma, Metasesarma, Metaplax, Portunus, Thalamita, Scylla, Charybdis, Mictyris Epixanthus, Pilumnus, Baptozias and Cardisoma.

Crabs of Mangroves (Many of assorted Colour)

In holes of Creeks	*In Deep Clayey soil*
– Scylla serrata	– Metaplex crenulata
Clayey soil	In mixed Sandy/Clayey soil
(Colourful ones)	– Uca lacteal, U. dussumere
Uca sp. Ocypoda sp.	U. tetragonon, U. vocuns
Dotilla sp. Sesarma sp.	
Grapsus sp. and Thelmeta sp., etc.	
At various sites	
– Ocypode	

Hermit Crabs

- **Intertidal sp.** **Clibanarius longitars**

Uca sp., Ocypoda sp., Dotilla sp.

Grapsus sp., Sesarma sp.,

Thelmeta Sp.

Important genera of Crustacea

– Acetes	– Coenobita
– Caridia	– Portunus
– Palaemon	– Scylla
– Thalsasina	– Thelamita
– Metaplax	– Balanus
– Sesarma	– Penaeus
– Uca	– Metapenaeopsis collected from the shell of Terebralia, Purpura, Achinata.
• Supra-littoral sp.	– Caenobita cavipes, collected from shell of Nerita, Cerithidea, Cerithium Polynices, Clypeomorus

These are a few genera and species.

6

Wildlife Parks and Sanctuaries

This chapter is a representation in part of the work recorded in the Encyclopaedia of Parks and Sanctuaries in Andaman and Nicobar island (1991). As the author's works done in 1987 have been reproduced in detail in the encyclopaedia, the author, has taken the liberty to reproduce those in this treatise.

List of National Parks and Wildlife Sanctuaries

Name of Park and Sanctuary	Area in ha.
National Parks	
Marine National Park	28150.00
Middle Button Island National Park	64.75
Mount Harriett National Park	4662.00
North Button Island National Park	44.00
Saddle Peak National Park	3254.88
South Button Island National Park	3.80
Total Parks Area	36179.43
Sanctuaries	
Arial Island Sanctuary (Ariel)**	5.00
Bamboo Island Sanctuary	5.00

(contd.)

Name of Park and Sanctuary	Area in Ha.
Barren Island Sanctuary	810.00
Battimaly Island Sanctuary	223.00
Belle Island Sanctuary	8.00
Benelt Island Sanctuary (Bennet)*	346.00
Bingham Island Sanctuary	8.00
Blister Island Sanctuary	26.00
Bluff Island Sanctuary	114.00
Bondoville Island Sanctuary (Boudevillue)*	255.00
Brush Island Sanctuary	933.00
Chanel Island Sanctuary (Channel)*	13.00
Cinque Island Sanctuary	951.00
Clyde Island Sanctuary (Clyd)**	54.00
Cone Island Sanctuary	65.00
Curlew Island Sanctuary	3.00
Curlew (B.P.) Island Sanctuary	16.00
Defence Island Sanctuary	1049.00
Dot Island Sanctuary	18.00
Dottrel Island Sanctuary	13.00
Duncan Island Sanctuary	73.00
East Island Sanctuary	611.00
East or Ingls Island Sanctuary	355.00
Egg Island Sanctuary	5.00
Elat Island Sanctuary (Fiat)*	936.00
Entrance Island Sanctuary	96.00
Gander Island Sanctuary	5.00
Girjan Island Sanctuary	16.00
Goose Island Sanctuary	1.00
Hump Island Sanctuary	47.00
Interview Island Sanctuary	13300.00
James Island Sanctuary	210.00
Jungle Island Sanctuary	52.00

Name of Park and Sanctuary	Area in Ha.
Kwangtung Island Sanctuary	57.00
Kyd Island Sanctuary	800.00
Landfall Island Sanctuary	2948.00
Latouche Island Sanctuary	96.00
Mangrove Island Sanctuary	39.00
Mask Island Sanctuary	78.00
Mayo Island Sanctuary	10.00
Megapode Island Sanctuary	12.50
Montogemery Island Sanctuary (Montgomery)**	21.00
Narcondam Island Sanctuary	68120
North brother Island Sanctuary	75.00
North Island Sanctuary	49.00
North reef Island Sanctuary	348.00
Oliver Island Sanctuary	16.00
Orchid Island Sanctuary	10.00
Ox Island Sanctuary	13.00
Oyster Island Sanctuary	8.00
Oyster Island Sanctuary	21.00
Paget Island Sanctuary	736.00
Parkinson Island Sanctuary	34.00
Passage Island Sanctuary	62.00
Patric Island Sanctuary (Petrie)**	13.00
Peacock Island Sanctuary (Pocock)*	62.00
Pitman Island Sanctuary Petman)**	137.00
Point Island Sanctuary	307.00
Potanma Island Sanctuary	16.00
Ranger Island Sanctuary	426.00
Reef Island Sanctuary	174.00
Roper Island Sanctuary	146.00
Ross Island Sanctuary	101.00
Rowe Island Sanctuary	11.00

(contd.)

Name of Park and Sanctuary	Area in Ha.
Salt Water Crocodile Sanctuary (Lohabarrack)**	222.00
Sandy Island Sanctuary	158.00
Sea Serpent Island Sanctuary	78.00
Shark Island Sanctuary	60.00
Shearme Island Sanctuary	785.00
Sir huge rose Island Sanctuary	1.00
Sisters Island Sanctuary	36.00
Snake Island Sanctuary	73.00
Snake Island Sanctuary	3.00
South Brother Island Sanctuary	124.00
South Reef Island Sanctuary	117.00
South Sentinal Island Sanctuary (Sentine)**	161.00
Spike Island-1 Sanctuary (Speke)**	42.00
Spike Island-2 Sanctuary	1170.00
Stoat Island Sanctuary	44.00
Surat Island Sanctuary	31.00
Swamp Island Sanctuary	409.00
Table (Delgarno) Island Sanctuary	229.00
Table (Excelstor) Island Sanctuary	169.00
Talabaicha Island Sanctuary (Talakaicha)**	321.00
Temple Island Sanctuary	104.00
Tillongchang Island Sanctuary	1683.00
Tree Island Sanctuary	3.00
Triby Island Sanctuary	96.00
Tuft Island Sanctuary	29.00
Turtle Island Sanctuary	39.00
West Island Sanctuary	640.00
Wharf Island Sanctuary	11.00
White cliff Island Sanctuary	47.00

MARINE NATIONAL PARK

This is one of the three marine parks in India, the other two being in Gujarat and Tamil Nadu. Stretching over 15 major islands and several islets of the Libyrinth Island group, its boundaries run sometimes along the coast, sometimes inland. The park encompasses an enchanting stretch of marine waters, lushly vegetated islands, vast coral reefs, and sparkling silver beaches. The marine fauna is immensely rich, and includes four (and possibly a fifth - see Fauna below) species of sea turtles, the dugong, and the saltwater crocodile, the flora is also extremely diverse, including among the largest remaining stretches of protected mangroves in India. The Park is separated by a small stretch of open sea from the salt water Crocodile sanctuary in the North.

Its proximity to Port Blair, its assessibility to both Indian and foreign tourists, and a concentration of settlements along its borders have resulted in considerable recent disturbance to this valuable area.

Some of the islands in the park are also known by their vernacular or local names: Lamba Pahar for Red skin Island, China Pahar for Boat Island, New Nariyal for Malay Island, Goltikri for Jolly Boys Islands, and Laltikri for Rifleman Island.

Flora

The major forest types include Andaman Tropical Evergreen Forest 1A/C2, Littoral Forest 4A/L1 and Mangrove Forest (Tidal Swamp Forest) 4B/TS2. There is no information on threatened species of flora in the Park.

A list of trees and other vegetation appears at the end of the sheet.

Fauna

Mammals [Q1, Khan undated, FD undated]

Bat, Dobson's Horsehoe	Deer, Spotted
Boar, Indian Wild	Dolphin, Common
Civet, Himalayan Palm	Fox, Flying

Lists of birds, reptiles and invertebrates are annexed at the end of the sheet.

The Green sea turtle and Leathery turtle are locally threatened due to poaching and egg collection. The Whitebellied sea eagle is also reported to be locally threatened [Q1, fv.].

The Crown-of-throns starfish Acanthaster planci, a predator of corals, is believed to be over populated in some parts of the park, and is causing considerable localized damage to the reefs (Wood 1989).

MOUNT HARRIETT NATIONAL PARK

Containing some of the highest peaks of the Andaman Islands, the Mount Harriett range is covered with dense evergreen and semi-evergreen forests. Having been left largely untouched, these forests are in prime condition, though growing settlements around the Park threaten them. While detailed studies have not yet been undertaken in the area, preliminary assessments indicate a rich diversity of floral and faunal life. The Park is also one of the few wildlife protected area in Andamans which is easily accessible. The area was earlier a Reserved Forest, notified as such on April 17, 1963.

There is a proposal to add portions of the North Andaman Reserve Forest to the west and south of the park (Het, Pers. comm. 1991). This would include Mount Harrientt Park itself, which seems to have been left out by mistake in the original notification (Khan, Pers, comm. 1991).

SADDLE PEAK NATIONAL PARK

Saddle Peak, at 737 metres above sea level, is the highest point in the A&N Islands. Shaped like a double-humped saddle, the park runs north to south along the eastern coast of North Andaman Island.

The one kilometer flat stretch between the sea and the foot of the peak has unfortunately been encroached upon, in recent years, by families from the Indian mainland. This is causing increasing destruction of the forests.

INTERVIEW ISLAND SANCTUARY

While approaching the island from the sea one is struck by the windblown appearance of the forests, especially in the north-west of the island. Obviously the Island marks the confluence of very strong winds which have permanently shaped the tree-tops.

Near the beach, in the south of the Island, there is a perennial freshwater pool inside a shallow cave, Legend has it that the pool has never been fathomed. Certainly efforts at finding the bottom with a ship's plumbline were not successful. In the cavern around the pool hundreds of Whitebellied swiftlets Collocalia esculenta have made their nests. The slightest disturbance sends these birds flying in and out of the sun light, looking very much like flashes of fire.

This island was contracted out for timber extraction in the fifties to a Calcutta firm, which started operations, laying tracks for its haulage coaches, and bringing in elephants from the mainland. Reportedly this operation ran into trouble and was abandoned, along with the elephants, who have since turned feral. Interview is now the largest island sanctuary in A&N.

36.41 ha teak was planted in the period 1956-63. Thinning operations were proposed to be carried out over this area from 1984, but there is no information if this was ever done (WP). 4-5 ha of Lagerstroemia hypoleuca were planted in the early 1970s (fv).

MEGAPODE ISLAND SANCTUARY

India's southernmost wildlife sanctuary, this is one of the last strongholds of the greatly endangered Megapode. It is a tiny island just off the south-western coast of Great Nicobar Island, thickly vegetated and ringed by sponge and coral beds. Unfortunately it is far easier to approach than most of the other island sanctuaries of Andaman and Nicobar, as one can walk across from the main Island during low tide. It is therefore susceptible to greater human interference. The Megapode population is reported to be threatened due to poaching and egg collection by local people.

North Button Island National Park, Middle Button Island National Park, South Button Island National Park.

These tiny islands, located off the south-eastern shore of Middle Andaman Island, form an arc in the northern part of Ritchie's Archipelago. They are reported to contain significant undisturbed breeding habitat for sea turtles as well as the endemic Andaman greyrumnped swiftlet (*Collocalia fuciphaga inexpectata*) (Chaudhuri 1987). No detailed ecological study seems to have been done of these islands.

NARCONDAM ISLAND SANCTUARY

This remote island sanctuary forms a stepping stone in the long chain of submerged and semi-submerged hills which extends from Burma and includes the Andaman and Nicobar Islands. Heavily forested, the Island is believed to be a craterless volcano long extinct indicating that the volcano may well have been recorded as active in human memory (Abdulali 1971). Its uniqueness lies in its population of the Narcondam hornbill, (Rhyticeros narcondami), a species found nowhere else in the world, but thriving here. The Island is difficult to reach without a good seafaring vessel, which is perhaps fortunate in that it restricts human disturbance.

Common and Scientific names of Mammals, Birds, Reptiles and Buterflies reported from National Parks and Sanctuaries in A&N Islands

[This contains a listing of only those species which have been recorded in national parks and sanctuaries in A&N]

Mammal

Bat, Andaman Horseshoe
Bat, Dobson's Horseshoe
Bat, Lesser Shortnosed Fruit
Boar, Indian Wild
Civet, Himalayan Palm
Deer, Barking or Muntjac
Deer, Spotted or Chital
Dog, Domestic (Feral)
Dolphin, Common
Elephant, Indian (Feral)
Flying Fox
Flying Fox (Andaman)
Flying Fox (Narcondam Small)
Flying Fox (Nicobar)
Flying Fox, Malayan Large
Goat, Domestic (Feral)
Rat, Brown
Rat, Common House
Shrew, Andaman Island Spiny
Shrew, Nicobar Spiny

Birds

Baza, Indian Blackcrested
Bee-eater, Bluetaiuled
Bittern, Yellow
Bluebird, Fairy
Bulbul, Redvented
Bee-eater, Chestnutheaded
Bittern, Tiger
Kingfisher, Whitecollared
Kite, Pariah
Koel

Crake, Andaman Banded
Crow, Jungle
Crow-pheasant
Cuckoo, Emerald
Cuckoo, Himalayan
Cuckoo, Indian
Cuckoo, Small
Cuckoo, Violet
Dove, Emerald
Dove, Red Turtle
Dove, Rufous Turtle
Dove, Spotted
Drongo, Andaman
Drongo, Ashy
Drongo, Great Racket-tailed
Drongo, Lesser Racket-tailed
Eagle, Andaman Dark Serpent
Eagle, Crested Serpent
Eagle, Whitebellied Sea
Egret, Large
Egret, Smaller
Falcon, Peregrine
Flycatcher, Brown
Flycatcher, Redbreasted
Flycatcher, Spotted
Harrier, Marsh
Harrier, Pale
Hawk-eagle, Crested
Hawk-owl, Andaman Brown
Heron, Chinese Pond
Lorikeet, Indian
Megapode
Minivet, Scarlet
Mayna, Common
Mayna, Hill
Mayna, Whiteheaded
Nightjar, Indian Jungle
Nightjar, Longtailed
Oriole, Blacknaped
Owl, Andaman Scops
Parakeet, Alexandrine
Parakeet, Redbreasted
Parakeet, Redcheeked
Petrel, Duskyvented Storm
Pigeon, Andaman Wood
Pigeon, Green Imperial
Pigeon, Greyfronted Green
Pigeon, Nicobar
Pigeon, Pied Imperial
Pipit, Redthroated
Plover, Great Stone
Rail, Bluebreasted Banded
Redshank, Common
Roller, Broadbilled
Sandpiper, Common
Sandpiper, Green
Shikra
Shrike, Brown
Snipe, Great
Starling, Glossy

(contd.)

Heron, Grey
Heron, Reef
Hornbill, Narcondam
Kingfisher, Blackcapped
Kingfisher, Blue-eared
Kingfisher, Common
Kingfisher, Ruddy
Kingfisher, Storkbilled
Kingfisher, Threetoed
Teal, Cotton
Teal, Grey
Teal, Lesser Whistling
Tern, Brownwinged
Thrush, Dark
Thrush, Siberian-Ground
Tree Pie, Andaman
Tropic-bird, Longtailed
Turnstone
Wagtail, Forest
Wagtail, Grey
Warbler, Ducky Leaf
Sunbird, Olivebacked
Swallow
Swallow-shrike, Whiterumped
Swift, Large Brownthroated
Spinetai
Swift
Swiflet, Andaman Greyrumped
Swiftlet, Himalayan
Swiftlet, Whitebellied
Warbler, Palefooted Bush
Warbler, Palelegged Leaf
Warbler, Thickbilled
Warbler, Yellowbrowed Leaf
Whimbrel
Whistler, Mangrove
White-eye
Woodpecker, Darjeeling Pied
Woodpecker, Fulvousbreasted Pied
Woodpecker, Indian Great Black

Reptiles

Crocodile, Estuarine or Salt-Water Crocodile
Gecko, Banded
Gecko, Dwarf
Gecko, Emerald
Lizard, Green Forest
Monitor, Water
Skink, Tytler's
Snake, Amphibious
Snake, Colubrine Amphibious Sea
Snake, flying
Turtle, Green
Turtle, Hawksbill
Turtle, Leathery
Turtle, Loggerhead
Turtle, Olive Ridley

Butterflies

Birdwing, Common
Clubtail, Andaman
Clubtail, Common
Helen, Andaman
Jay, Great
Jay, Tailed
Mime, Common
Mormon, Andaman
Mormon, Common
Mormon, Great
Rose, Common
Rose, Crimson
Sapphire, Purple
Sunbeam, Burmese
Swordtail, Fivebar

7

Biosphere Reserve in Great Nicobar

Great Nicobar Biosphere Reserve should be considered to be a paradise for biographers (phyto and zoo). This reserve has reflected a special status of its own due to the treasure of endemic flora and fauna, mixed biota of South and Southeast Asia and a forested land and not yet molested by India's growing population.

This reserve will be ecoeducation tourists' esteemed study area, which is virtually a museum and a living laboratory for the researchers for the days to come, where the communication infrastructural facilities will improve.

OBJECTIVES OF BIOSPHERE RESERVE

- Identification and demarcation of the unique ecosystem and its role for international network in global ecosystem conservation.
- Ecologically compatible economic development (Eco-development) of the intertidal zone.
- Research, training and monitoring etc.

BACKGROUND INFORMATION

Biosphere Reserves

The Indian Man and Biosphere (MAB) Programme, modeled

after the UNESCO (MAB) Programme, envisages the conservation of representative ecosystems which are genetically and ecologically rich. Such ecosystems are designated Biosphere Reserves, with the "objectives of

a. conserving the diversity and integrity of plants and animals within natural ecosystems,

b. safeguarding genetic diversity of species on which their continuing evolution depends,

c. providing areas for ecological and environmental research, and

d. providing facilities for education and training."

[Ministry of Environment and Forests, 1989c]

The model followed internationally also envisages the designation of several zones in each Biosphere.

Reserve

a. ***Natural or Core Zone***: Managed for minimum human interference to serve as a baseline for the biological region; research, educational and training activities are carefully controlled and must be non-manipulative.

b. ***Manipulative or Buffer Zone***: Managed for research, education and training activities, and manipulative methods and techniques are permitted. Traditional activities including timber extraction, hunting, fishing, and grazing are permitted in a controlled manner.

c. ***Reclamation or Restoration Zone***: Managed to study and reclaim lands and natural resources where heavy natural or human-caused alteration have been interrupted or where species have become totally extinct.

d. ***Stable Cultural Zone***: Managed to protect and study ongoing culture and land use practices which are in harmony with the environment. Local residents and their activities be strictly controlled. [Ministry of Environment and Forests, 1989c].

So far, the designation of Biosphere Reserves is not done under any law. It is now proposed to introduce a clause allowing for this, in the Wildlife (Protection) Act of 1972. Alternatively, in the long run, a specific law on this may be considered [Ministry of Environment and Forests, 1989c]. At the moment, management of the Reserves is in the hands of the concerned State Government, while the Central Government extends full financial assistance for approve activities, technical expertise and know-how, and detailed guidelines for management.

The Great Nicobar Biosphere Reserve

On January 6, 1989, the Ministry of Environment and Forests, Government of India, designated a major part of Great Nicobar Island as a Biosphere Reserve, by Government Order No. J. 22010/14/89-CSC [Ministry of Environment and Forests, 1989c]. The Reserve extends, in two parts, over 88,500 ha. (885 sq.km) or about 85% of the Island. The area covered is not part of, nor does it contain, any existing national park or sanctuary, though the designation of the biosphere reserve as a park or sanctuary has been suggested by the Wildlife Institute of India.

Great Nicobar Island, the southern most island of this archipelago and in fact the southern most land piece of India, is situated between 6°45′ and 7°15′N latitudes and 93°38′ and 93°55′E longitudes. The island lies about 482 km south of Port Blair and 145 km north of the northern tip of Sumatra. The island is about 55 km long between Murray point in the north to Pygmalion Point (now called Indira Point) in the south. It has a width of about 30 km in the north but the island narrows down to only about 3 km in the southern tip. The total geographical area approximates to 1045 sq.km.

NEED FOR BIOSPHERE RESERVE

The following is a summary of the justification for establishment of a Biosphere Reserve in this island.

(1) The island is situated at a phytogeographically strategic point.

(2) About 85% of primary forest areas of the island are still virgin and rich in species content.

(3) Endowed with immense genetic resources of wild plant species such as timber trees, fruit trees, ornamental plants, medicinal plants, etc. the island harbours rich genetic germplasm resources.

(4) About 11% of the vascular flora (so far surveyed) are endemic to these islands.

(5) About 30% of the flora are extra-Indian i.e. occurring (elsewhere) only in S.E. Asian countries.

(6) Details of faunistic survey certainly show endemism among animals in the islands, as well as their interrelationship with plants.

(7) About 30 species of plants are rare among the known flora, and are endangered and confined to one or few localities each on this island.

(8) Cyathea albo-setacea, a characteristic tree fern and Phalaenopsis speciosa, a beautiful ornamental orchid are endemics, and found only in this adjacent islands.

(9) Similarly, some endemic birds are characteristic of the island ecosystem (e.g., the Megapode-Megapodus freycinet nicobariensis).

(10) The Shompens, an aboriginal tribe with very limited population, are a source of much valuable information regarding the usefulness of many unknown species.

(11) Impact of external forces, especially tourism, is practically nil at present.

(12) Politically the island is at a strategic position, and grave danger looms over the flora or fauna from accelerated developmental activities in the near future.

ZONATION OF THE RESERVE

Entire northern part of the island starting from the Alexandra river (west coast) to Chengappa Bay (eastcoast) i.e. between 7°N and 7°20′N latitude, and 93°37′E and 93°6′E longitudes as Core Zone I. (Area-520 sq.km as Core Zone-I and 90 sq.km as its Buffer Zone). The southern part between the two hilly ranges (Sahni range and Mani range) including the river Galathea (6°60′ N and 7°N latitude and 93°37′E and 93°56′E longitudes as Core Zone-II. (Area-185 sq.km as Core Zone-II and 90 sq.km as its Buffer Zone). (Please see map for zonation).

Special Features of Great Nicobar Biosphere Reserve (Area: 1045 sq.km):

• Distance from	:	Port-blair-482 km. Indonesia (Sumatra) - 138 km.
• Flora	:	Fern-77 sp., Gymnospem - 3 sp. Dicot-438 Sp., Monocot - 130 sp.
• Fauna		Nicobar Megapode, Nicobar pigeon Nicobar pig, Crab eating macaque, Man-eating crocodile.
• Sweet water rivers	:	Galathia, Nanda, Alexander, Degmar, Amrit Kaur, Jubulia
• Southernmost point	:	Indira point
• Rehabilitation	:	Ex Army men
• Endemism	:	Very many species of plants and animals.
• Hills	:	Chaturvedi Range & Mount Jhuliar Peak.
• Beautiful plants	:	Cyathia sp. (Tree fern), Lindsaya sp (a fern). Cycas rumphii (a rare plant)
• Bay	:	Galathia, Campbell, Casuarina, Trinket.

Rare and Endangered Animals

- Andaman pig
- Nicobar pig
- Dugong dugon
- Large Andaman Teal
- Andaman Falcon
- Crab-eating Macaque
- Man-eating estuarine crocodile
- Nicobar megapode
- Nicobar pigeon

Great Nicobar Island is the southern-most island in the Andaman-Nicobar-Archipelago and is also the southern-most point of land within the Indian Union. Located between 6°45′ and 7°15′ North latitudes and 93° 40′ and 93° 55′ East longitude, this island is situated 510 km. South of Port Blair. It covers an area of 1,04,454 hectares. Till the 1960's the island was inhabited by only less than hundred Shompens and a small number of coastal Nicobarese. Exploration of the island for locating settlements commenced only in the late 1950's.

Understanding of tropical soils and rain forest ecosystems was poor at that time and the availability of plentiful fresh water and the luxurience of natural rain forests led to wrong conclusions on the suitability of the land for intensive cultivation and settlement.

The current view on conversion of tropical evergreen forests to agricultural land is expressed in the following summary on the tropical rain forests from the Earthscar paper of the International Institute for Environment and Development.

"The tropical rain forests usually go to make way for land hungry farmers. But farming is often the worst use of tropical rain forest soils. Though lush, most hold nutrients in the vegetation, not in the soil. When cleared most tropical rain forest soils are virtually sterile.

Intact tropical rain forests are a warehouse of the genetic diversity needed to provide crop varieties to feed the world. They support known and as yet unknown medicinal, food and fuel plants. They protect watersheds, preventing floods, erosion and siltation."

As forests disappear, so do the tribal people who live in them. A World Bank report suggests that only indigenous forest dwellers know how to derive a sustained living from the tropical rain forests. Outsiders with their short-term profit-goals are usually doomed to failure.

UNIQUE LOWLAND TROPICAL RAIN FORESTS

The Great Nicobar Island is of unique importance to India from environmental point of view. It contains the only sizeable untouched lowland evergreen forest in the country and is an excellent example of an ecosystem in which climate, soil, vegetation and fauna are components in an extremely complex equilibrium. When one of these components, the primary forest is destroyed, whether partially or completely, by felling and burning, the other components undergo an equally violent change. The lowland evergreen forests of the Indo-Malayan region are the world's richest and at the same time the least studied of the terrestrial ecosystems. In Great Nicobar Island, the evergreen forests, the coral reefs and other land-marine ecosystems interact with each other intimately. Consequently, the variety of habitats and naturally adapted species along these land-water junctions are unique.

These islands have a large number of endemic plant species and also some faunal species. Giant coconut crab, the crab eating macaque, the Nicobar Megapode and Nicobar pegions are some examples. It is presumed much wealth of the islands are not yet fully known.

8

Grand Effort to Save the Tribals

Areas Marked for different tribes are:

- In Great Nicobar for the Shompens (119 sq.km)
- In Strait Island for the Andamanese (0.6 sq.km)
- In the Sentinel Island for the Sentinalese (60.00 sq.km.)
- In Little Andamans for the Onges (252 sq.km.)
- In Middle & South Andamans for the Jarawas (638.86 sq. km.)

In addition to the 100 national parks and sanctuaries, substantial areas in the islands are demarcated as Tribal Reserves, under the Andaman and Nicobar Islands (Protection of Aboriginal Tribes) Regulations of 1956. Their primary function is to safeguard the interests of the vulnerable tribal communities of Andaman and Nicobar Islands. The wildlife is not strictly protected, since traditional hunting, trapping, and other such activities of the tribals continue. The chances of this having led to any serious wildlife decline are remote, given the low populations and simple technologies of these tribes. The only exception to this may be in Great Nicobar Island, where the Nicobarese hunt the Megapode (Megapodus freycinet) and collect its eggs.

The following are the designated tribal reserves [Chana, Pers. comm., 1989]:

1. 11,900 ha on Great Nicobar Island, for the Shompens
2. 25,200 ha on Little Andaman Island, for the Onges
3. 6,000 ha on North Sentinal Island, for the Sentinalese
4. 60 ha on Strait Island, for the Andamanese
5. 63,886 ha on the western part of Middle and South Andaman Islands, for the Jarawas.

There is also a move to declare large areas in the islands as Biosphere Reserves. A large part of the Great Nicobar Island has already been designated as the Great Nicobar Biosphere Reserve while the entire North Andaman Island along with several surrounding islands is proposed to be given the status of North Andaman Biosphere Reserve [Ministry of Environment and Forests 1989b].

National and international anthropologists have studied the ethical history of these tribes and have postulated their theories about the origin and behavioural pattern of these tribes. Yet many inner details of their society and life style are not known and the ecoeducationists may find a virgin field for research.

Population

Onge - 97 Jarawa - 200-300

Sentenalese - 50 Andamanese - 26

Shompen - 223

The origins of the 6 tribes of A&N are unclear. There is a distinctive racial difference between the Andaman tribes and their counterparts in Nicobar. The former are Negrito, while the latter are Mongoloid. This, along with other indications, has produced certain theories regarding their respective origins. It is believed that the Andaman tribals had reached the islands ". . . as early as 670 A.D. . . ." [Whitaker 1985], and arrived by boat from South East Asia since they ". . . show a considerable affinity to the Semangs of Malaysia and the Aetas of Philippines. . . .". [Witaker 1985]. The Nicobarese ". . . must have migrated sometime before the Christian era from what was then a part of India, the land now occupied by the Burmese, Talaings (Mons),

Shans and Malays. They are offshoots of some Mongoloid race." [Singh 1978]. The origin of Shompens is not known ". . . but it looks as though the Shompens have a Malayan strain which, in all probability is mixed with the Dravidian" [Singh 1978]. The origin of Shompens is not known ". . . but it looks as though the Shompens have a Malayan strain which, in all probability is mixed with the Dravidian" [Singh 1978]. The Andaman tribes are primarily hunter-gatherers, while the Nicobar tribes are mainly horticulturalists and herders.

The six tribal groups have had varied contact with outsiders, and assimilation into the Indian mainstream. On the one hand are the Nicobarese, who have integrated well, while on the other hand are the Jarawas and the Sentinalese, who still violently resist most efforts at contact. Somewhere along the middle range are the Shompens, who avoid contact but are not particularly hostile, and the Andamanese and Onge, who have accepted the presence of outsiders.

A look at the population figures of all the A and N tribes, except the Nicobarese, and perhaps the Harawas, shows a downward trend till recently, which is in sharp contrast to the population figures of settlers. These tribals are mostly, in fact, just barely holding on (see below).

The horror story of the A and N tribes, can be traced back to the arrival of the first settlers along with "modern civilization" at Port Blair, though in later years a number of other factors were also contributive.

It was in the late 18th Century that the sea faring nations of the West made their contact with the Islands, mostly to replenish their supply of drinking water. The British...made a first attempt at establishing a settlement at Port Cornwallis (North Andamans) in 1788. This was abortive. A second attempt made in 1858 in Port Blair. Land was cleared for a prison and a penal colony set up. [Saldanha 1987].

THE ANDAMAN TRIBES

After the setting up of the penal colony in Port Blair, several skirmishes between the Andamanese and the British took place, with many killings on both sides. Finally, friendly relations with the tribals were established. However, new killers emerged in the form of diseases brought in by the settlers, to which the tribals had no resistance, like pneumonia, syphilis, and measles. This along with a shrinkage of their habitat, due to the ever increasing influx of settlers, resulted in a rapid erosion in the number of the Andamanese. Today, from a population of around 5,000 in 1857, only 32 Andamanese (including three non-tribals, a lady and her two children from previous marriage, who have joined a tribal family) survive on Strait Island [Saldanha 1987]. However, the population has stabilized somewhat in the last three decades; having gone down to a precarious 19 in 1961, it has slowly climbed up subsequently [Verma 1989].

The first contact of the Onge with outsiders in 1867, on their former stronghold of Little Andaman Island, resulted in 70 of them being killed. However, after friendly relations had been established, the same story of succumbing to diseases and being pushed back by the arrival of settlers was repeated. The Onge have now been relegated to a reserve at Dugong Creek, having to share their former territories with about 7000 setters and 700 Nicobarese (Saldanha 1987). From a population of 672 in 1901, their population has gone down to a mere 98 in 1989 [Verma 1989]. Their numbers, like that of the Andamanese, have stabilised over the last two decades (Verma 1989).

The Jarawas, who are today confined to a tribal reserve in the western part of Middle Andaman and South Andaman, as well as the Sentinalese, located on the North Sentinel island, are the tribes which have till recently remained hostile to outsiders.

The hostility of the Jarawas has been given varied explanations [Awaradi 1990]. One of the first theories is that they became so because of the activities of slave traders from the Malayan islands, centuries ago. Another theory holds that the Jarawas resented the British attempts to befriend them with the help of the Great Andamanese, with whom the Jarawas had relations of enmity. Yet another hypotheses is that the Jarawas became hostile after seeing the decimation of other Negritos, like the Andamanese, at the hands of the British. In addition, due to their hostile position, the British are reported to have undertaken many punitive missions against them [Whitaker 1985]. These tribals also suffered at the hands of the Japanese during the Second World War, when the A and N islands were controlled by the latter for around two years. Many Jarawas were then captured, tortured, and killed, ostensibly as part of the efforts of the Japanese to curb the underground resistance organized by the British in A and N (Kothari 1989). Due to the above mentioned factors, the population of the Jarawas continued to decline till about the 1940's. However, though reliable estimates are obviously impossible to obtain, their number is reported to have increased, from abound 50 in 1951 to 200 in 1985 (Saldanha 1987; the 1981 census also reported a population of 200). If this is true, this is probably due to their isolation from the outside world.

A small population of Jarawas, near the main settlement areas in South and Middle Andaman Islands, has been responding to these efforts since the first friendly contact in 1974, but the tribe as a whole remains hostile to casual efforts at contact and intrusion into their territory (Awaradi 1990; Pandit 1991). Incidents of Jarawa, attacks on people who have intruded into their territory or settled on its eastern fringes continue; a total of 133 such incidents have been recorded between 1947 and 1988 [Awaradi 1990]. A new threat to the existence of the Jarawas has emerged in the form of the Diglipur-Port Blaid Andaman Trunk Road, which is mostly bordering the reserve, and to some points even cuts through it. This road has not only reduced the area of the Jarawa reservation, but has also made it vulnerable to greater disturbance [Saldanha 1987]. The road, which is now acknowledged even by the A and N Administration to be an error [IDA 1991], also has the potential of aggravating the conflicts between the Jarawas and travellers on it. Of late there has been a friendly trend among the Jarawas who are accepting the material help extended to them and have ceased to be hostile to the local residents.

The Sentinalese did not, till recently, have any contact with outsiders. However, in 1967 two different teams landed, in close succession, on the North Sentinal Island [fv]. Though no direct contact was made with the Sentinalese, they were spotted and presents were left for them in their huts. Subsequently, another expedition's visit to the island reportedly resulted in a confrontation. The Sentinalese apparently became hostile after this, and were not met until anthropologists were able to establish friendly contact in January and February 1991 [Pandit 1991].

Approximately 80 of these tribals occupy the North Sentinal Island (Verma 1989, citing the 1981 census), and their being left alone at present is probably the best that can be done for them and their long term survival. This has been suggested recently by S.A. Awaradi, former Director of Tribal Welfare, A and N Administration (Awaradi 1990).

THE NICOBAR TRIBES

The Shompens, who inhabit and greatly depend on the forests of the central uplands of Great Nicobar Island, currently face a threat from the construction of a road from Campbell Bay to Kopenheat, horizontally bifurcating the island [Saldanha 1987; fv]. Plans to commence timber extraction operations on Great Nicobar also pose a potential threat to their survival. Perhaps the recent declaration of Great Nicobar as a Biosphere Reserve (see Appendix 8 for details) will help to protect the life-support systems of the Shompens. As in the case of the Andamanese and the Sentinalese, no precise estimate of their population is a available. The census of 1901 put their numbers at 348; more recent estimates by R.K. Tiwari, a social worker who has lived with the "Shompens for years, put the population at 212 in 1981, declining to 149 in 1989 (Verma 1989).

As mentioned earlier, the Nicobarese form the only tribe in A and N which has been able to adapt to modern civilization. This may partly be due to the efforts of many missionaries over the last few centuries. At the moment these tribals are greatly concentrated on Car Nicobar Island, and are mainly engaged in plantation activities and horticulture. 18% of the Nicobarese are literate, while 12% go to school (Saldanha 1987).

It is clear that, with the exception of the Nicobarese, all the indigenous tribal communities of A and N are perilously vulnerable, or close to extinction. This is by far one of the most important issues facing the island's administrators today.

THE KARENS

Parts of the coastal areas of North Andaman Island, especially around Mayabander, are inhabited by the Karens. These are descendants of a Burmese tribe, whose members were brought as labourers by the British. An American Baptist Mission in Burma sent 45 families in 1925, on the request of the British government (Awaradi 1990). After Independence, these people were settled as agriculturists on the Middle Andaman Island. Since then, the Karens have acquired an excellent knowledge of the rainforest and the seas, becoming expert hunters and fisherfolk [fv]. The single keel 'Karen boat', strong and mechanized, has become a famous 'brand' in the islands [Awaradi 1990]. These people regularly poach in Interview Island, and fish in the nearly waters. Temporary settlements are made on the Island, or the abandoned timber camp once used by the P.C. Roy Company is utilized for staying the night in search of Spotted deer, Indian wild boar, and turtles [fv]. Weapons used include guns and spears, and dogs. Reportedly the Karens are assisted by some police stationed on Interview Island, in return for which the police get regular supplies of meat and other items [fv]. Also, it is reported that the Karens occasionally guide other hunters on the Island [fv].

9

Management Strategy

FOREST MANAGEMENT

It will be evident from the historical background of the Department that management was oriented towards exploitation of commercial species for earning revenue and to obtain regeneration of a few commercial species only. Owing to the pressure of 36 woodbased industries for timber and various development work there has been colossal demand for timber and entire forestry machinery has been geared to cause the supply. Their demand for more timber as increasing over the years (2.05 lakhs cum a year). The Forest Department apprehended such demand is a threat to the protection of forest in the island particularly in the vicinity of the existing hamlets of settlements. Extraction of timber using elephants, tractors and locomotives in the slopes have caused appreciable erosion.

Although the Forest Department has fixed up 4085 ha or 1,85,000 cum as annual yield there is apprehension that over felling is being done because, (i) area assessment has no sound foundation; (ii) there is no growth data to rely upon and the fixation of rotation and conversion period is far from correct. Canes and other minor forest produce are also being over-exploited. The Forest Department extracted 1,38,126 cum of timber during 1985-86, but wastage of logs and timber was also high.

Natural regeneration of 41,975 ha. has so far been achieved. Clear felling of forests was done in the past to raise Teak, Padauk and Pyinma plantations, but this method has been discontinued (plantation since raised over 24,860 ha.). Many conservators have severely criticized the method of selection felling and the entire canopy lifting operation for obtaining regeneration of commercial species. They thought that soil got much sunlight to dry up; burning damaged regeneration and made area drier; seedlings of evergreen species failed to germinate and then establish. So the subject was still debatable. There was no cell in the Department of Forests to monitor the regeneration operation and evaluate the forestry and environmental aspects of natural regeneration process. McVean found natural seedlings on forest floor but observed that litter and humus was lacking on forest floor.

At present the Department of Forests is said to have shifted emphasis from the productivity of commercial timber alone to forest conservation.

The mangroves have been exploited heavily because of high calorific value of wood as fuelwood. The present method of management prescribes felling in belts/patches. This may have to be discontinued. The present requirement as stipulated by the Forest Department is about 6150 cum per year specifically meant for Electricity Department for power production.

The plantation and Forest Development Corporation is working as lumbering Department diluting the functions of silviculture, protectionary and regulatory functions. Oil palm plantation has been raised over 1593 ha. The projected target was 2400 ha. But at the moment the raising of plantation has been banned in Sarthankhari in South Andaman and at Kalchal.

CONSERVATION STRATEGY

The area of National Parks and Wild Life Sanctuaries had been planned to be increased to 2,000 sq km or about 24 per cent of the geographical area of the territory. An area of 74,000 ha has been notified to be Jarwa reserve and another 16100 ha as Sentenelese reserve. The conservation status of this 79,990 ha is not known.

Analysing the fact presented in this and earlier chapters it is felt that conservation of small islands and other small areas would be inadequate to give full protection to the flora and fauna of this island in entirety. Conservation of total area of 2000 sq.km. of Sanctuaries and National Parks may initially satisfy many and boost up the percentage of National Parks and Wild Life Sanctuary areas in relation to geographical area but most of the plots are very small (except probably Interview Island) to give flora and fauna adequate space to interact, regenerate and survive as a Community and as individuals.

FOREST MANAGEMENT VIS-À-VIS ENVIRONMENTAL CHANGE

The forests of Andaman are one of the richest repository of tropical forest resource in India. An area of about 1150 sq.km. has been deforested since earlier settlement.

C.G. Rodgers, C.J. Lyal, Sir David Prain, CE Perkinson were the foresters and botanists that explored the forests and studied their resources. Regular plan was prepared by F.H. Todd in 1905 for North Andaman Forests which entailed improvement felling, adoption of strict fire control measures and seed dibbling in blanks. Bonington's working plan (1914) suggested felling of over mature trees. This was followed by concentrated regeneration felling. Thereafter B.S. Chengappa (1952), J.C. Vermah (1960), ACD Cruz (1970-71) worked in the forests all over the north, middle and south Andamans.

In fact Padauk and Teak were raised at selected spots. By 1983, an area of 5046 acres was planted up but only a marginal success was achieved. In 1970's palm oil was introduced in little Andaman by felling of virgin forests. Many consider the raising of teak plantation in Andaman as a resource mismanagement.

In a tropical evergreen forest micro-climate in the forests has to be preserved. And alteration of it can only be counter-productive leading to forest degradation. Opening up the canopy beyond certain limit, allow more sunlight to penetrate, more heat created in forest floor and allow torrential rain drops to hit the soil destroying top soil and disbalancing nutrient cycle. Removal of a small percentage tree (as prescribed in Working Plans) created large gaps all over (as the trees had vast crown cover) and therefore humid atmosphere below top canopy changed. In felling operation, logging and extraction of timber trees countless regeneration of trees, shrubs and herbs were damaged. The movement of heavy vehicles, skidders and caterpillar tractors destroyed vast expanse of undergrowth and soil exposed to atmospheric acid especially to heavy rain. As a result hydrological characteristics of the worked areas changed vastly. Everywhere over felling of trees, damaged trees and cyclone affected trees were noticed especially in Little and North Andamans. The very nature of evergreen forests have changed due to felling of plywood and matchwood species over vast areas.

The foresters have already exploited the best areas so far timber is concerned. It is difficult to find a glimpse of once existing giant tropical evergreen trees in the Andaman after walking through scores of kilometers.

Geologically and geomorphologically Andaman and Nicobar islands have certain special character which have a considerable bearing on human existence there. These are natural climax forests. The islands are essentially unconsolidated marine sediments composed of mud, clay and sand with only sand stone, mud stone or shale as harden sulastratum material. There is no crystalline rock and, therefore, weathering and erosion is extremely high. In sandy or coralline substratum percolation is extremely rapid with little run-off while in muddy or clay substratum percolation is very little and run-off is very high.

Destruction of forests have increased the humidity of coastal water which might effect coral reefs in continenta shelf. The drainage in the hills is very rapid and the substratum can hardly retain much water. So there is dearth of ground water sources. The apparant dampness and lashness of these islands were due to the unbroken climax forest which insulated the ground surface and retained humidity.

It has been an ecological disaster caused to Andaman forests in settling lakhs of agriculturists all over the islands. Forestry is yet to guarantee the protection cf soil, water and maintenance of suitable agroclimate.

RICH ENDEMIC SPECIES NEED PROTECTION

The inland forests and coastal waters of Great Nicobar Island are already known to harbour the largest concentration of endangered plants found nowhere else within this country, while several species occur nowhere else in the world. Rare endemic animals such as the Giant coconut robber crab, the crab-eating macaque, the Nicobar Megapod, the Nicobar pigeon etc. occur only in the Great Nicobar Island. This island has in addition currently least disturbed fresh water streams with crocodiles and water monitors. Three or four species of marine turtles nest along its beaches, perhaps the only area in India where so far they have remained undisturbed. Only a minute fraction of the biotic wealth of this island is as yet known to science.

Vulnerability of tropical rain forest ecosystem

The tropical evergreen forests are the world's most complex and integrated ecosystems wherein by mutual harmony and maximum diversity, the highest productivity over unit area is maintained in the entire biotic world. They are however highly vulnerable to any serious disturbance, particularly to fragmentation. It is far more damaging to locate a number of scattered pockets of human activity within rather than a single large concentration.

Great Nicobar Island's unsuitability for development

Ecologically the Great Nicobar Island is highly unsuitable for development due to a variety of reasons. Among these are the heavy, well distributed rainfall with practically no dry season, very steep undulating terrain, very fragile ecosystems which rapidly deteriorates with any interference, lack of any hard rock for construction activity and proneness to earthquakes. The existing east-west road in the island from Campbell Bay to Kopanheat is an example of what could happen due to building activity. Unconsolidated and steep clay hills tend to slip if they are undercut and exposed. More than half the length of the east-west road has become unusable and the vegetation along its alignment has been degraded; unchecked erosion is now carrying silt to the Galathee river drainage damaging more extension forests than directly destroyed by road building. Maintaining the road will be prohibitively expensive, apart from accentuating the fragmentation of Great Nicobar island's forest into two halves. The stability of the road for effective use is doubtful too.

Coastal road alignment

Because of difficulties of terrain, roads in Great Nicobar Island will have to be aligned along the coast, where road construction of boulders and sand result in sea erosion. The removal of coastal vegetation in turn damages inland forests and crops. Besides, there is also the indirect loss of coastal and marine biotic communities due to siltation etc. The forty mile long North-South road from Campbell Bay to Galathea river mouth provides examples of all these consequences.

Ex-servicemen's Rehabilitation

In Great Nicobar island, for strategic reasons, rehabilitation of ex-servicemen from many states of India was initiated in 1969 and so far 7,055 acres of forests have been cleared to settle 330 families numbering 2,116 people at considerable expenses. About 300 acres of this area is flat land along the south-eastern coast, ecologically these tropical islands are not suitable for agriculture. They have an intrinsically poor soil made further vulnerable to the leaching action of torrential rains - about 3500 mm per year - spread throughout the year. In the long run it would be uneconomical to cultivate these lands as cost of maintaining soil fertility and to combat pests and pathogens would be prohibitively high. Even now since individual families were given 11–14 acres of land each, loss of the land is being put to sub-optimal use due to lack of adequate manpower. According to data provided by the local administration about 1500 acres are currently lying unutilized. Conditions of life are difficult and have forced 52 settlers to return to the mainland. The Rehabilitation Department has already spent Rs. 63.83 lakhs directly on these settlements, and individual assistance to the extent of about Rs. 25,000/- has also been extended. Substantial overhead expenditure on establishment and on institutional finance has also been incurred. In spite of this colossal investment, these settlements are far from being economically viable. In the process, substantial invaluable primary evergreen forests have been lost. The basic cause for this unhappy situation is the ecological unsuitability of this island for any kind of agricultural settlement.

Defence requirement

The Defence requirements so far projected involve clearing of 4000 acres of previously untouched forest areas and for quarrying purposes at least one more area will have to be deforested. From Campbell Bay to Galathea Bay, a distance of at least 38–42 km of coastal ecosystems have been damaged already. At Pygmalian point in India, further damage is taking place. Serious attention need, therefore, be given to the possibility of locating as much of the area for defence needs as possible from within the already cleared rehabilitation areas in order to minimize further damage. Even if substantial compensation has to be paid and the settlers given the option to choose their future, either to return to the mainland or to be translocated within the Andaman Group of islands in revenue areas which are still unallocated, from the environmental conservation point of view, this would be a highly desirable step.

[P.S. Last six paragraphs are modified version of Dr. S. Nair]

10

Development Activities *vis-à-vis* Conservation Problems

So far the qualities and floral resources of the Islands have been presented in a very broad outline. A few such qualities and associated problems are:

QUALITIES

i. Vast floral and faunal resources,

ii. Interesting geographical situation,

iii. Bonanza of biogeographical affinities,

iv. Poor status of soil, devoid of nutrients and extremely poor in water retention capacity,

v. High degree of endemism of floral and faunal species which are endangered because of a very limited space for each.

PROBLEMS

i. Rapid increase in population (from about 30,000 in 1951 to 1,88,000 in 1981) resulting in pressure on land,

ii. Too many wood based industries (36 numbers) utilizing a heavy quantity of timber for which heavy felling has been made,

iii. Soil deterioration owning to introduction of coconut and oil palm plantation,

iv. Vast developmental activities involving many Government Departments, settlement of large number of people and creation of ever increasing infrastructural facilities,

v. Settlement of unspecified number of fishermen,

vi. Lack of field based research.

Although the Govt. of Andamans have already taken adequate steps it still have to take further numerous steps for improving the all out ecological status. Of late

a. There has been a total moratorium of all large scale development projects and on further settlement of people as it is thought that the present population has exceeded the bearing capacity of soil and further settlement will destroy the biological resources.

b. There has been a drastic reduction in the number of woodbased industries. This will minimize the quantum of timber supply to these industries and obviously less felling of forests will be involved. The forest and Plantation Development Corporation has recently started two match factories and one saw mill in Little Andaman and at Mayabandar. It is not known on what basis it was done.

c. There has been complete suspension of further felling of forests in Little Andamans for planting oil palm.

The following points need serious attention for effective conservation of the biogeographical resources.

(a) Should an island forests which have acquired balance after millions of years isolation, be suddenly subject to massive modification?

(b) Should there be more Govt. and Non-Govt. offices, research centers, institutions etc. as those will invite more people require more fuelwood, more timber, more land and create other related problems?

(c) Should there be a levish use of expensive timber in building construction?

(d) Should the settlers still enjoy gift of timber and other forest produce?

(e) Should this island provide employment for persons coming from mainland?

(f) Should the forest be managed on commercial line-exclusively?

These are intrinsic limiting factors in Andaman and Nicobar environment which will delimit the development plannings and prove such effort as unproductive. The forests of Andaman and Nicobar Islands are only the relict of 'Tropical Rain Forests' in India and there is already indication of species loss. Therefore, these forests are also to be preserved to provide aesthetic and recreational facilities for the millions of Indian tourists and to shelter myriads of vertebrate and invertebrate animals besides plants, to explore for scientific and limited purpose of economy.

Since 1965 several expert groups engaged by the Govt. of India have visited the islands primarily for determining (i) suitability areas for settlement of refugees, (ii) integrated resources development, (iii) accelerated agricultural development. The problem of deforestation, however, appealed the then Prime Minister of India, Smt. Indira Gandhi and in her initiative a Multidisciplinary Team of officials visited the island in 1975. This team and another land use committee set up later stressed for collection of data and preparation of land use and problems of conservation. It was Mc Vean the I.U.C.N. expert who visited the island in 1976 recorded for the first time the serious loss of forest resources and exploitation beyond the carrying capacity of land exacerbated by unchecked immigration. He thought the islands tropical forest, coastal and marine ecosystem could be saved, and suggested complete suspension of forest clearance. This was not accepted by the Govt. of India. As a follow up, the Govt. of India appointed a team of Administrators who visited the island and recommended (i) preparation of base industries; and (ii) creating of a Corporation viz. The forest and Plantation Development Corporation.

It is paradoxical that the barring Dr. Mc Vean neither the foresters, nor other scientists nor the administrators, thought of taking drastic steps on conservation of fragile soil and vegetation cover.

None except Dr. Mc Vean raised his voice against settlement of people after clearing forests. By now more than 1150 sq.km has been diverted for non forestry purposes. Encroached area is 2784 ha. The Department of Forests suspect more encroachment in interior area which has remained undetected because of inadequacy of the protection personnel. Repatriates from Burma and Sri Lanka have been accommodated in addition to largescale settlement of erstwhile East Pakistan refugees. In Little Andaman 28645 ha were marked for rehabilitation and 200 ha for nicobarise settlement and another 164 ha for defence. In Great Nicobar 7055 ha were marked for ex-servicemen. In 1980-81 1463 fishermen were engaged in fishing and rehabilitated 126 families. If marine fishes have to be exploited heavily and aquaculture to be developed, a vast population has to be rehabilitated. This indicates a danger to the island and, therefore, needs judicious handling.

Provision for land for future requirement of Defence and O.N.G.C. (which may expand enormously should oil be struck) has to be made. Besides these, there would be Administrative personnel, with enormous staff size, Forest Department Staff and skeleton staff of other essential Departments only. Unless the Govt. establishments are restricted to the minimum the population pressure and their infrastructural requirements will nullify all conservation efforts.

Problems centering the Islands

- Soil erosion
- Population expansion
- Poor capacity of soil to hold water
- Conservation of ornithofauna
- Too much expansion of wood based industries
- Conservation of flora
- Limited agricultural production
- Conservation of fauna
- Development programmes (lack of environmental protection issues)

- Lack of research data
- Unsuitable agricultural land

DEVELOPMENT ACTIVITIES

The A and N islands are ecologically and socio-culturally very fragile. The development activities in these islands must be undertaken stricktly keeping this in mind. The tragedy however is that most development activities undertaken in the past (as well as many of those which are proposed), have been (or are likely to be) mostly destructive from the ecological and social point of view. The policy of settling people in large numbers, and lack of discouragement to unplanned entry of people from the mainland, has been coupled with encouragement of activities which are out of harmony with the environment, and a direct threat to the tribal people of these islands. A spurt in activities like agriculture, plantations, logging and related activities like timber based industries, construction, and unplanned or unregulated tourism have already damaged the ecological and cultural richness of the islands. A number of reports by expert committees have taken the view that any further large-scale damage is not advisable [Anon 1986; Anon 1988].

Future development of the islands is therefore a great challenge. Two urgent considerations are uppermost here: one, the need to make the islands as self-sufficient in food and other essential items as possible; and second, the need to ensure sustainable employment to all inhabitants. The sectors from which these two objectives can be met are agriculture, industry, fisheries, forestry, tourism, and services. Each of these has its own set of potentialities and problems.

(a) Agriculture

The Planning Commission estimates that at most 5 persons per hectare can be supported by agriculture in A and N, and "considering the topography, and need for biosphere reserves, tribal and forest reserves, it is unlikely that the total (potential) agricultural land, including crops and plantations would be much more than 50,000 ha." [Anon 1986]. This implies, that the present population of A and N is already over the maximum agricultural carrying capacity of the islands. Also, about 46,000 ha are already under cultivation and plantations [Anon 1986]. Indeed, a number of factors militate against substantial expansion of this sector; no unforested flat land left, poor soils, lack of ground water, high rate of soil erosion on exposed slopes, pests like the Giant snail, danger from exotic weeds, and the negative impact of agricultural practices like fertilizer and pesticide use.

However, it is possible that the spread of more appropriate land uses and technology, and encouragement of fisheries (see below) could augment the food production, reducing the need to import from the mainland.

Cash cropping in the islands is currently dominated by trees like Coconut, Red oil palm, Betelnut, Rubber, etc. There have been a number of problems with these plantations, including low yields, ecological damage, pest attack, wastage in harvesting operations, and the like [Anon 1986; Anon 1988]. While no great increase in plantation area is advisable, since it will involve clearance of natural forest, there is plenty of scope for optimizing yields from the existing plantations, and for intercropping with non-tree crops in agricultural fields. Additionally, crops like coffee, cocoa, clove, nutmeg, pepper, and cinnamon can be grown in moderation within forest areas, without significant ecological damage [Anon 1986]. There is also a potential for setting up ancilliary units, provided they are environmentally sound, for processing and using by-products.

(b) Industry

As in the case of agriculture, there is not much potential for generating employment in this sector. Most of the industrial units at present are forest-based, and will have to be converted or shut down when timber operations are phased out (see below, Forestry). Already, the WIMCO management is contemplating shutting down its match factory at Port Blair [Mohanty 1991]. Ecological and energy considerations make any expansion of large industrial units inadvisable. The viable options for the islands then include small and medium-scale industries based on fisheries and other marine resources, plantations, animal husbandry, and similar activities [Anon 1986].

(c) Fisheries

This is widely perceived as the sector with the greatest employment potential. It is estimated that as against a present annual catch of about 6500 tonnes, the potential yields can be of the order of 50,000 to 8,20,000 tonnes (the reasons for this wide range of estimates are unclear) [Anon 1988]. A number of measures have therefore been suggested by various expert committees, including motorisation of traditional boats, introduction of deep sea trawling, training in efficient fish-catching techniques, and extension of marketing facilities [Anon 1986; Anon 1988]. Industries ancillary to the fisheries sector like canning and processing, or boat building, could also provide employment.

However, several doubts have been raised about the wisdom of going in for a sudden and unplanned expansion of fisheries; the danger of over exploitation (especially present when rushing into an area with rich potentials), the potential conflicts between traditional fisherfolk and trawler owners (so apparent in most of India's coastal areas), the risk of marine pollution and damage to coral reefs, and other similar problems. There is also the ethical question to be tackled, which has made hunting of most terrestrial creatures an unacceptable employment generator.

Finally there is the danger of simplistic assessments of 'carrying capacity' of the islands based on the fisheries potential For instance, one proposal asserts that the 6 lakh sq. km. exclusive economic zone (EEZ) around the islands is capable of supporting a population of 6 lakh people, based on the formula that each square kilometer of the EEZ is capable of supporting one person (Kothari 1989). The basis on which these figures were calculated, is not at all clear, nor is it clear if other factors (energy, habitation and freshwater needs of such a large population, etc.) have been considered.

(d) Forestry

Since, the IDA has taken a decision to phase out timber and other extraction activities in natural forest areas, this sector will have very little employment potential. There is also not much scope for largescale plantation activity; already, fresh teak plantation has been given up, and no natural forest areas are to be cleared for raising any other plantations [Sinha 1991]. However, replanting of degraded areas, and other such environmental restoration programmes can be expanded in the future.

(e) Tourism

At first glance, the A&N Islands are eminently suited for rapid and large-scale expansion of tourism. Repeated administrations have tried to push ambitious schemes to promote the visit of tourists, especially foreigners. However, the necessity of heavily subsidizing tourists (in terms of expensive services like freshwater, electricity, transportation, food provisions, etc.) makes this sector economically problematic. In addition, uncontrolled or inappropriate tourism can have severe ecological and cultural repercussions, as are already being felt in parts of the islands. Expansion of this sector is therefore possible only in a limited way.

(f) Others

There is likely to be some expansion of employment possibilities in other sectors, such as services, trade, and commerce. One proposal which has generated a lot of controversy, is that of a Free Trade Zone in Great Nicobar. This could have potentially disastrous effects on the ecological sustainability of the islands, and cultural survival of its tribals. Fortunately this proposal seems to have been shelved for the time being.

It is clear that these islands cannot tolerate any further large scale expansion of population, or of industrial, agricultural, and forestry activities. At current trends of immigration and natural growth of the existing population, the islands will have over 4.15 lakh people by 2001. However, if immigration can be drastically curtailed, the population will be a more manageable 3.13 lakhs [Ministry of Labour 1988]. Any future developmental plan for A and N must be mindful of this. It must respect both the fragility of the area's natural ecosystems, as also the survival needs of its indigenous populations.

Development strategies for the A&N Islands are laid out by the Island Development Authority (IDA), set up in 1985. With the Prime Minister of India as its Chairperson, the IDA is serviced by the Environment and Forests Division of the Planning Commission, and has, as its members, various cabinet ministers, secretaries, and non-governmental experts. Increasingly, the IDA itself has considered, the ecological and social fragility of these islands as being of the highest priority, a view that is endorsed by a range of non-governmental groups and individuals in the fields of wildlife and environment, anthropology, social work, and health.

[**Source**: *Directory of National Parks and Sanctuaries 1991.*]

11

Check List of Faunal Resources

A check list of animals of A and N groups are compiled below [Source: Tikadar and Das, 1985 and Sesingh's National Parks and Sanctuaries of Andaman and Nicobaris (1991)].

Mammals

Common Name	*Common Name*
Bat (17 species)	Cat, Jungle
Civet, Andaman Masked Palm[2]	Deer, Barking
Deer, (Spotted) or Chital	Dolphin, Common
Dugong	Flying Fox (6 species)
Macaque (2 species)	Mouse, House
Pig (2 species)	Pipistrelle (2 species)
Rat (8 species)	Rat House (5 species)
Shrew (6 species)	Squirrel, Northern Palm 5
Vampire, Malay False	Whale (3 species)

Birds of Andaman and Nicobar Island. [Sources: Tikadar and Das (1985) and S. Sing's Directory of National Parks and Sanctuaries (1981)]

Common Name	*Common Name*
Baza, Andaman Blackcrested	Bee-eater, Vluetailed
Bere-eater, Andaman Chestnutheaded	Bittern, Chestnut
Bittern, Nicobar Tiger	Bittern, Yellow
Bluebird, Fairy	Bluethroat, Northern
Bulbul, Andaman Blackheaded	Bulbul, Andaman Redwhiskered
Bulbul, Nicobar	Bunting, Uttle
Bunting, Yellowbreasted	Chat, Siberian Blue
Chat, East Siberian Collared Bush	Cock, Water
Crake, Andaman Banded	Crake, Eastern Baillons
Crow, Eastern Jungle	Crow, Pheasant, Andaman
Cuckoo (8 species)	Cuckoo-Dove, Nicobar
Cuckoo-Dove, Nicobar	Cuckoo-Shrike (3 species)
Curlew (2 species)	Dove (3 species)
Drongo (6 species)	Duck, Spotbill
Eagle (6 species)	Egret (4 species)
Falcon (2 species)	Flowerpecker, Andaman 3
Flycatcher (7 species)	Godwit, Bartailed
Goshawk, Horsfield's	Greenshank
Harrier (3 species)	Hawk Eagle, Andaman Crested
Hawk Owl (3 species)	Heron (7 species)
Hoopoe, Tibetan	Hornbill, Narcondam
Kestrel, East Himalayan	Kestrel European
Kingfisher (10 species)	Kite (2 species)
Kite, Pariah	Knot, Eastern

(contd.)

Common Name	Common Name
Koel, Andaman	Lapwing, Greyheaded
Lorikeet, Indian	Magpie Robin, Andaman
Megapode, North Nicobar	Megapode, South Nicobar
Minivet, Andaman Scarlet	Minivet, Ashy
Minivet, Eastern Small	Moorhen, Indian Purple
Moorhen, Malay	Munia, Andaman Whitebacked
Munia, Nicobar Whitebacked	Myna (6 species)
Nightjar (2 species)	Noddy, Whitecapped
Oriole, (3 species)	Osprey
Owl (4 species)	Parakeet (4 species)
Partridge, South Indian Grey	Pastor, Rosy
Peafowl, Common	Pelican, Spottedbilled or Grey
Petrel, Wilson's Storm	Pie, Andaman Tree
Pigeon (6 species)	Pipit (3 species)
Pitta, Nicobar Hooded	Plover (8 species)
Pratincole, Collared	Quail, (2 species)
Rail, Andaman Bluebreasted	Banded Redshank, Common
Roller, Andaman Broadbilled	Sandpiper, (5 species)
Shama, Andaman	Shikra, Car Nicobar
Shikra, Katchal	Shrike, Brown
Shrike, Phillippine	Snipe (4 species)
Sparrow (3 species)	Stare, Andaman Glossy
Stint, (4 species)	Sunbird (4 species)
Swallow (3 species)	Swift (2 species)
Swiftlet, Andaman Greyrumped	Swiftled, Himalayan
Swiftlet, Whitebellied	Teal (4 species)
Tern (8 species)	Thrush (6 species)
Tropic-bird Redtailed	Tropic-bird White

Common Name	*Common Name*
Turnstone	Wagtail (6 species)
Warbler (13 species)	Waterhen, Andamans Whitebreasted
Whimbrel	Whimbrel, Eastern
Whistler, Mangrove	White-eye, Nicobar
Woodpecker, Andaman Black	Woodpecker, Andaman
Woodpecker, Spottedbreasted Pied	

Lizards and Reptiles of Andaman & Nicobar Island Source: Tikadar and Das (1985 and S. Singh's Directory of National Park and Sanctuaries 1991)

Common Name	*Common Name*
Bronze Back (4 species)	Cobra (2 species)
Crocodile, Estuarine	Gecko (11 species)
Krait (3 species)	Kukri, Andaman Banded
Lizard (8 species)	Monitor, Andaman Water
Python, Reticulated	Skink (15 species)
Snake (25 species)	Tokay, Asian
Tortoise, Malayan Box	Trinket, Green
Trinket, Redtailed	Trinket, Yellowstriped
Turtle (6 species)	Viper (5 species)
Endangered animal, birds and reptiles	
Dolphin, Common	Dugong or Sea cow
Macaque, Crab-eating	Pig, Andaman Wild
Whale, Blue	Whale, Spern
Birds	
Eagle, Whitebellied Sea	Falcon, Peregrine
Hornbill, Narcondam	Megapode

(contd.)

Common Name	*Common Name*
Osprey	Pigeon, Nicobar
Teal, Grey or Andaman	
Reptiles	
Crocodile, Estuarine	Monitor, Andaman Water
Python, Reticulated	Turtle, Green
Turtle, Hawksbill	Turtle, Leathery
Turtle, Olive Ridley	

12

Study in Marine Ecosystem

OCEANOGRAPHY

The Bay of Bengal (including Arabian sea and Indian ocean, which are closely interlinked) has been identified by the UNESCO as one of the world's 49 large ecosystems for proper preservation of its resources and sustained development. It is located in the monsoon belt that covers the three 'Golden Triangle' sites and bridges Sundarbans and the Andamans.

Two presentations one on marine plants and another on marine animals record broadly the marine resources of this ecosystem and how these resources cater to the need of the people of the area. For eco-education tourists this marine ecosystem provides an immense study materials [Sources: UNESCO Workshop; at Chennai Feb - 2002].

Some of the following findings were recorded in the UNESCO Workshop; which highlighted varieties of resources:

- 42 Genera of Andaman coral have been reported.
- 14 species of sea-grasses have been found in the area.
- Coastal population generates heavy quantity of solid wastes.
- Over-exploitation of resources through coral mining.

- Destructive finishing practices degrade the coral ecosystem.
- Destruction of mangroves over extensive areas.
- Huge industrial wastes discharged in the sea causes extensive water pollution. Marine plants and animals are eco-education tourists' wide field of bio-ecological rendezvous.
- Physical oceanography.
- Marine floor
- Ocean circulation
- Nutrient cycle in the sea
- Waves
- Biological oceanography (Plankton, Nekton, Benthos)
- Tides
- Marine sediments
- Mineral wealth
- Deep sea sediments
- Continental Shelf
- Marine geotechnology
- Continental slope
- Marine archaeology
- Marine fauna.

MARINE ANIMALS

A broad overview of some selected groups.

Between Sundarban and Southern most point of Nicobar islands lie in the Bay of Bengal and Andaman sea flaked by the Arabian sea and Indian ocean. This presentation on marine plants and animals covers a very broad area beyond these bays, seas and ocean. The ocean as a whole in one worldwide marine ecosystem.

Mystery shrouds around actions and interactions of marine animals flourishing in a vast and complex aquatic ecosystem affected by various factors such as physical, chemical, mechanical and biological. In this short non-technical presentation, the importance of a very few selected groups of marine biota, such as zooplankton, zoobenthos, corals, turtles, fishes and mammals have been discussed in brief outlines to show how each of these groups interact with phyto and zooplankton and also with each other and maintain the balance of food web, nutrients and energy level in marine environment. The relationship among phytoplankton, zooplankton and various organisms has been drawn in a very brief outline restricting the discussions on marine animals only. This presentation has been made for amateur visitors.

MARINE ENVIRONMENT - A COMPLEX ECOSYSTEM

Marine water is a vast ecosystem. It is not just a 'drowned land'. The science of marine fauna is a vast subject of oceanography and embraces a large field of science, viz., physics, chemistry, geology, hydrogeology and several other disciplines besides, biology.

About 97 per cent of the total one million splices (approx) of animals that exist in the world today belong to the invertebrate group (without backbone); the remaining 3 per cent belong to vertebrates (with backbone) such as fishes, amphibians, reptiles, birds and mammals. In marine water all vertebrate groups except the fishes are poorly developed. Invertebrate fauna, however, dominate over wide range of habitat. [Source: Marine Life by J.P. George *et al.* 1979].

OCEAN - A BIOLOGICAL LABORATORY

There are bewildering varieties and diversities in marine organisms which baffle human knowledge and imagination. The role of tide, ocean current, wave turbulance, planktonic flora and fauna, benthic (bottom) and pelagic (water body) fauna, sea weed, meadow, kelp forests in shapping the intricate life process of marine mammals (reptiles, fishes and other macro animals) is not yet fully known.

The principal marine life styles are planktonic (floating), nektonic (swimming) and benthic (bottom soil), squids, prawns, fish, turtles, sea snakes, seals, sea-cows and whales etc. are terminal consumers of the sea mostly being carnivorous (few are herbivores).

MARINE ZOOPLANKTON (ALSO PHYTOPLANKTON) PROVIDES FOOD FOR MANY MARINE ANIMALS

Marine zooplankton include a very wide varieties of organisms. Crustacea, however, form the bulk zooplankton organism (70%) and the predominant group being copepoda. This often outnumbers all the other animals group and are represented by many species. Another important group is Euphausiacea (Euphausia superba occurs in enormous numbers near the Antartica). Of this Euphausia, 'Krill' survives, and thereby the predatory group of giant whales feed on them.

The Zooplankton also include varieties of protozoans and molluscs (Pteropoda). In the Andaman Sea and Southeast Asia copepods, molluscan larvae, choetognaths, decapod larvae, Lucifer etc. predominant. The zooplankton include both herbivorous and carnivorous types.

The herbivores feed on phytoplankton. The planktonic herbivores are mainly copepods, euphausids, cladocerans, mysis, the cosomatours, petropods and the unocordales. Planktonic carnivores include medusac, ctenophores, copepods, chaetognaths, polychaetes, hyperid, amphipods etc. [Source: Elements of Marine Ecology by R.V. Tait 1980].

Nectonic organisms are actively swimming animals of the marine ecosystem. Crabs, squids, fish, whales etc. fall in this group. Some of these animals undergo long horizontal migrations over thousands of kilometers, others migrate periodically to great ocean depth. Benthic organisms inhabit the bottom as they are mud dwellers (isopods, sponge hydroids, bristle slars, sea-urchins, shrimps etc.)

SOME MARINE MAMMALS *vis-a-vis* PLANKTONIC FOOD

The cetaceans include the whales, porpoises, dolphins, seals and walruses. Whales are a source of meat, fats and oils, hormones, such as insulin, and chemicals. They exist at all levels of ocean food chain. The blue whale mainly devours small red dish shrimp called 'Krill', while the formidable killer whale feed on salmon, seals, and sharks. Dugongs (sea-cow) and anatees are plant eating aquatic mammals. They are common shallow coastal wasters of little Andamans. Dugongs are slow moving solitary creatures of the Indo-pacific region.

TURTLE MYSTERY IN MARINE WATER

"Protected by their shells perfectly adapted to a life spent mostly at sea; the turtles watched the dinosaurs come and go, and have been serenely going about their business ever since." [Source: A Agardy of W.W.L.] But skill with which they have survived the millennia may not be adequate to save them from human greed for food and commercial values.

These animals have been in existence for 80 million years. All the eight marine species are now threatened with extinction. Of these, the Kemp's ridley turtle (Lepidochelys Kempii) is on the verge of complete extinction.

The logger head (Caretta caretta) is the common turtle and occurs in tropical water. The green turtle (Chelonia mydas) also has world wide distribution and is common in Sunderban and Andaman waters. Indian water has hawkbill turtle (Eretmochelys imbricata) ridley turtle (Lepidochelys olivacea), logger head turtle (Caretta caretta) and Leather back turtle (Dermochelys coriaria). The olive ridely turtle is being cultured on a small scale in West Bengal. Large scale annihilation of this species has been reported from Orissa coast. Australian flat back turtle (Natalor depressa) is confined to eastern Indian ocean and in the South Pacific.

Green turtle feed primarily on seagrass and algae; logger heads on crustacea; hawkbill on sponges and other invertebrates on coral reefs; ridleys on fishes, shrimp and crabs, leather back on soft-bodied jelly fish. So the turtle have specific preference and there is no clash on the food stock. However, in acute scarcity conditions they become omnivorous. The leatherback turtles which feed on jelly fish has to travel thousands of kilometers in search for food as it needs bulk supply to keep going its 600-1000 kg body weight.

CORAL REEFS ARE NATURE'S PARADISE

Coral reefs are naturalist's paradise where the water is clear, warm and the underwater land surface shape their intriguing forms among which swarm a Kaleidoscope of brightly coloured fish. The rash quantitites of limestone produced by corals, their high productivity, the richness of species have been ever the curiosity of the biologists.

Coral reefs are sometimes regarded as equivalent to tropical rain forests for their species diversity. The life history of corals is typical of sedentary animals. As India and other countries of South East Asia have extensive areas under coral reefs there seems to be immense scope for in depth research.

Mariculture potential of Indian sub-region

Mariculture is gradually assuming importance for increasing seafood production. There are numerous bays, lagoons, creeks and inlets with varying depth and wide range of substrata which are optimum for mariculture. In the present trend of population explosion in India the development of agriculture alone would not be enough to meet with food requirement; development and conservation of marine resources would be a solution in this regard.

Thailand has the World's biggest restaurant - 'The Royal Dragon' spread over 3.34 hectares which is primarily a seafood restaurant. Another such restaurant "Thai Place" own 2.72 hectares has been functioning in the country. The Royal Dragon will have 1200 employees, the menu offers more than 1000 dishes and will require 1000 kg of seafood every day. As such the mangrove forests and continental shelf areas have to be protected which are rich in oysters, dams, cockles, scallops, mussels, other bivalves squids, octopus molluscs, crustaceans such as lobster, penaeid prawn, crabs etc. and many other edible invertebrates and also fin fishes and reptiles to provide human food.

EDIBLE INVERTEBRATES

The species are green mussel (*Perna ulridis*), Squids, Octopus, Jelly fish (*Rhopilema esculenta*) Sea urchin (*Diadema setosum*); Sea cucumber (*Holothuria* 'sp), Turbu (*Turbo marmoratus*), Trochus (*Trochus niloticus*) are all edible species that occur in Southeast Asia.

The best known marine crustacea that form delicious food for man is *Penaeus monodon* (tiger prawn), *Scylla serrata*, an edible crab, is also a delicious food. Other edible crabs are miask crab, blue swimming crab etc. The Indian white prawn (*Penaeus indicus*), ciant banana (*Penaeus merguiensis*), Yellow prawn (*Metapenaeus brevicornis*) and rainbow prawn (*Parapene opsis hardwickii*) are also in great demand in the commercial market.

Metapenaeus sp., *Penaeus marguiensis* and *P. monodon* form very important food item in Thailand.

Six species of lobstar viz., *Panulirus homarus, P. ornatus, P. penicillatus, P. versicolor, P. polyphagus* and *P. longipe* are also in demand commercially. The giant clams or the holy water dams (*Tridaciia crocea, T. maxima* and *T. squamosa*) occur in Andaman water which are edible.

All these invertebrates have commercial market in the world and can be cultivated at suitable sites.

BIOLUMINESCENCE

Some marine organisms have the property of production of light (like the terrestrial glow worm). Dinoflagellates, Jelly fish, lantern fish, etc. have this property to glow emitting blue light. The light of some worms and ctenophores is blue green, whilst in one deep sea fish (*Echinostoma ctenobarba*) a large luminescent cheek organ flashes with a blue or pink light. *Noctiluca scintillans* generate phosphorescence in the luciferin-luciferace system).

OCEAN - THE WONDERLAND OF MARINE FISH

Melodious voice and innocent eyes of brilliantly coloured birds are divine, the strength, varied forms and behavioural pattern of mammals are magnificient, but marine fishes are natures masterpieces of creation in their gracefrul movement, assemblages of Kaleidoscope of brightly coloured fins, varied forms and adaptations intriguing life cycle which shades all other groups of animal world, terrestrial or marine.

The fish is cold-blooded aquatic vertebrates and is distinct from other air breathing animals, amphibians, reptiles, birds and mammals. The fish fauna maintain a delicate balance with the biological, chemical and physical factors of the aquatic environment.

DIVERSE FORMS AND RESOURCES

There are probably more than 20,000 different species of fish in existence today and 100 or more new forms seems to be discovered every year. (J.R. Norman, *A History of Fishes*, 1948).

About 14 per cent of total fish catch of the world is from fresh water, the balance 86 per cent is from marine. Of the marine fish, shell fish account for 10 per cent of this marine catch. The Pacific ocean yield about 5.5 per cent; the Atlantic ocean account for 39 per cent; the Indian ocean about 4 per cent; and 2 per cent form Mediterranean and Black Sea.

Typical consumers of marine plankton are herring, menhaden, sardines, and pilchards. Owing to plentiful food source in the ocean these fishes have high concentration.

Fish provides 23 per cent of the Worlds protein intake. Harring and Cod fishes are mostly harvested in the world followed by jacks, sauries, red fish, mackerel, tuna, flat fish and others. In southeast Asia Harring (giant herring, wolf herring, round herring, big eyed berring, five-spot herring, Chinese herring), Indian Tarpon, milk fish, sardine, Long-jawed Anchovylel, Cat fish, mullet, tiger fish, sand fish, hammer-headed shark, sword fish etc. are commonly found.

THE OCEAN - A VAST ECOSYSTEM

The oceans occupy 71 per cent of earth's surface area and form a major single ecological unit, the marine biome. The average depth is about 3700 metres and the maximum depth is about 11,000 metres. This vast water mass (1400 million cubic kilometers) support myriads of plants and animals. [*Source*: *Encylopedia of Astronomy*: Consultant C. Ronar]

The marine biome is the world's largest. It has immense Chasmdeep-Sea trenches - which is deeper than the highest land mountains. The ocean comprise one-single inter-communicating system with largescale currents and floating organisms.

Mightly currents mix, stir and circulate heat across the glob, regulating climate. Flood, drought and long term changes are ultimately controlled by dynamic, ever changing ocean. (The Statesman Oct. 8, 1995). Devastating cyclones, typhoons besides tide, tradewind monsoonic cloud formation all originate in the ocean. The vast water mass of the ocean have sheltered myriaids of species of plants and animals which have undergone evolution over the years and sustained the human kind since its creation. Yet only a little about the marine biota and the mystery of the ocean has so far been known.

OCEAN-PROVIDE FOOD

a. For Man

Ocean provide both plant and animal food (only plant food will be discussed in this article) though only a small quantity of plant comes from the sea. Various marine algae (macro) are taken as human food in Japan, England, Oceanic islands and Southeast Asian countries primarily; some such algae are - Red Lever (Porphyra umbilicalis), Green Lever (Ulva lactuca), Carragreen Moss (Chondrus crispus), Gigartia stellate and Pepperdulse (Laurencia pinnatifida). Carragreen grows all over the north Atlantic ocean. Several seaweeds have commercial value being the source of 'Alginic acid' which finds its use in the food industry as an emulsifier.

The principal source of 'Agar' of commerce is from macro sea algae e.g. Gelidium, Gracilaria, Pterocluda, Acanthopettis and Annfetlia. 'Agar' is used in bacteriological work in culturing some micro organisms. These are also used in canning fish, transporting cooked fish, sizing fabrics, making films, finishing leather products, producing high grade adhesive, cooking (soups, jellies etc.), manufacturing of ice-cream, icing, clarifying beer and wine, manufacturing cosmetics and in very many other industries.

Glacillaria confervoides - a red algae is cooked for food in Thailand; *Griffithsia corolline* is cooked for food in Vietnam. Several other such species are cooked and eaten in the Philippines of which *Porphyra* sp., *Codium* sp., *Enteromorpha* sp., *Sargassum* species are common.

b. Sea weeds (Sea grass, kelp and Algae) for marine animals

Some macroscopic members of the group Chlorophyta (green algae), Phaeophyta (brown algae) and Rhodophyta (Red algae) living in the ocean form food for selective groups of animals. These plants are visible to the naked eye; they are generally attached to solid substrate between and below the tide level. These groups also include angiosperms (flowering plants), which are sea-grasses.

The sea grasses of the meadows have vascular structure and root/rhizome system (as opposed to algae) generally inhabit near shore shallow water. These areas are highly productive, species - rich submarine meadows which provide ecologically important habitat for a number of commercial fin and shell fishes. The primary commercial value of seagrass meadows are the feeding habitat for commercially valuable fishes. Seagrass stabilizes sediments and reduce floor erosion. These plants produce useful biomass when associated with epiphytic algae and benthic (sea floor) macro algae. Sea grasses are marine, though a few extend upto inter-tidal zone. Even at a depth of 8000 m these plants have been found in ocean floor. Mats of leaves to the extent of 50 m diameter have been recorded in Florida coast. Common genera are Thalassia, Syringodium, Zostera etc. Sea-grass detritus is rich in micro organism such as bacteria, flagellates and ciliates. Epiphytic algae sticks to the leaves.

CLIMATIC CHANGE (TROUBLESOME TWINS EL NINO AND LA NINO) *vis-a-vis* PHYTOPLANKTON PRODUCTION

For reasons not clearly known the sea water in the west coast of Peru showed a rise of temperature by 10°C in 1980. This had a disastrous effect on the planktonic flora and fauna and fish population, besides bringing several other climatic disasters.

The name 'El Nino' (Christ child) was given to this phenomenon of warm water surge which is technically known as the El Nino Southern Oscillation (ENSO). It occurs when a reduction in the Southeast trade winds allow strong current of warm surface ocean water to sweep eastward across the Pacific and then down the South American Coast. This pushes the usual cold, northward flowing currents out to oceans preventing the upswelling of cold, nutrient rich deeper water and significantly inhibiting the biological productivity of coastal areas. Many sea birds starve and commercial fish catch affected, besides occurrence of torrential rain and even drought. It is feared that similar phenomena may affect the South Atlantic coast of Africa and the Indian Ocean, and that El Nino itself is at least partly responsible for longer term climatic changes and later America and World wide El Nino was made responsible for poor development of plankton, torrential rain in dry areas of California and Equador, fire devastation in New Guinea, and drought in South Africa, Australia, Indonesia and India.

[*Source*: *The Green Book 1991* by Stephen Pope *et al.*].

The phenomenon due to cold water in Pacific Ocean has been termed 'La Nino'—The girl.

[*Source*: *The Green House Effect 1989*, by Swert Boyle *et al.*].

These phenomenon may upset the entire balance of flora and fauna in the marine water. El Nino Phenomenon may occur as the oceanographers think every two to ten years).

PLANKTON AMELIORATE GREEN HOUSE EFFECT AND GLOBAL WARMING

The 'Green House Effect' and 'Global Warming' are serious concern of the world today. There has been heavy accumulation of carbon dioxide in the atmosphere, besides oxides of nitrogen, methane, chlorofloro carbon and ozone. Each year 4.8 to 5.8 billion tons of carbon is released in the atmosphere of which the earths atmosphere takes up approximately 2.9 billion tons (Carbon dioxide emissions are conveniently measured in terms of their carbon). The rest is removed by the plants, by natural process in carbon sink - places like ocean.

[*Source*: R.A. Sedjo, *Environment*, Vol-31, Jan. 1989].

The ocean is one of the great sink for carbon dioxide. Some 10 billion tons of carbondioxide are said to be fixed each year as a result of photosynthesis by plants (like phytoplankton and sea weed) in the ocean. The sea weed take about 10 per cent of it, and the rest by the plankton. Of the carbondioxide which disappear in the ocean, it has been thought that about half of it is dissolved in the water and another half is taken by plankton.

The Scientists in British Research Ship Challenger found that only about 30 per cent of the carbon dioxide dissolve into the water; as such the plankton must, therefore, by playing a more important role.

[*Source*: James Wilkinson, *Green Bust*, 1990].

Bibliography

SUNDARBANS

Annandale, N. 1922. *The marine element in the fauna of the Ganges,* Bijdr Dierk (Feest Num. Max Webers). 22: pp. 143-154.

Baidya. A. and A. Choudhury. 1984b. Distribution and abundance of Copepod in a tidal creek of Sagar Island, Sundarbans, West Bengal. *Environment & Ecology*. Vol. 2 (4): 00. 33-37.

Banerjee, R. 1985. Estuarine Crocodile. *J. Born. Nat. His. Soc.* pp. 214-215.

Blanford, W.T. 1891. *Fauna of British India: Mammalia.*

Blasco, F. 1975. *The Mangroves of India,* Inst. Francais de Pondichery. Trans. Sec. Sci. Tech. Vol 14: pp. 1-175.

Chakraborty, K. and A.B. Chaudhury, 1972. Honey and production and Behaviour potential of the honey-bee. *Science and Culture.*

Champion, H.G. and S.K. Seth. 1968. *A revised survey of forest types of India.* Government of India.

Chaudhuri, A.B. and K. Chakraborty. 1972. Observation on Tiger. *Cheetal.* Vol. 15(1).

Chaudhuri, A.B. and K. Chakraborty. 1973a. Wildlife biology of the Sundarbans: marine fish. *Science and Culture.* 39 (I): pp. 8-16.

Chaudhuri, A.B. and K. Chakraborty, 1973b. Breeding biology of birds of Sundarbans forests. *Cheetal.* Vol. 15(2): pp. 7-31.

Chaudhuri, A.B. and K. Chakraborty. 1973c. Wildlife biology of the Sundarbans forest: A study of the habitat of the tiger. *Bull. Bot. Soc. Bengal.* 23: pp. 63-66.

Chaudhuri, A.B. and K. Chakraborty. 1979. *Tiger and Mass. Proc. of Int. Symp.* 1979.

Chowdhury, M.K. and P. Sanyal. 1985a. Vegetation Dynamics of Mangroves – Proc. All India Symp. On Marine Plants Goa. Use of convulsion shock to control tiger Predation: *Tigerpaper.* Vol. XII: pp. 1-5.

Chowdhury, M.K. and P. Sanyal. 1985b. Man-eating behaviour of the Tigers of Sundarbans, West Bengal. *Environment and Ecology.* Vol. 3: pp. 243-248.

Curtis, S.J. 1933. *Working Plans for forests of Sundarbans Division,* Bengal Government Press, Vols. 1 and 2.

Department of Marine Science. 1987. *Report on Mangrove Ecosystem of Sundarbans.* University of Calcutta.

Department of Marine Science. 1992. *Mangrove ecosystem of Sudarbans in virgin and reclaimed Areas with special reference to productivity.* University of Calcutta.

Government of India, 1990. *Conservation of Mangroves in India.* Ministry of Environment and Forests, New Delhi.

IUCN, 1989. *Marine Protected Areas Needs in the South Asian Seas Region.* Volume 2: India.

Mandal, A.K. and R.K. Ghosh. 1989. *Sundarbans: A Socio-bioecological Study.* Bookland Private Ltd., Calcutta.

Mandal, A.K. and N.C. Nandi. 1989. Fauna of Sundarban mangrove Ecosystem, West Bengal, India. *Fauna of Conservation Areas 3.* Zoological Survey of India.

Mukherjee, A.K. 1969. Food habits of water birds of Sundarbans, 24-Parganas District, West Bengal, India. *I.J. Bombay Nat. Hist. Soc.* Vol. 66(2): pp. 345-360.

Mukherjee, B.B. and J. Mukherjee. 1978. Mangroves of Sundarbans. *Phytomorphologia.* Vol. 28 (2): pp. 177-192.

Naskar, K.R. and D.N. Guha Bakshi. 1982. Sundarbans - the world famous mangrove forests of the District 24-Parganas in West Bengal. *J. Econ. Tax. Botany.* 3: pp. 883-918.

Naskar, K.R. and D.N. Guha Bakshi. 1987. *Mangrove Swamps of the Sundarbans.* Naya Prakash, Calcutta.

Oldham, R.D. 1893. *A manual of Geology of India,* Calcutta.

Pargiter, F .E. 1934. *A revenue history of the Sundarbans from 1765 to 1870*. Alipore, B.G. Press.

Andaman

Abdulali, H. (1964). The Birds of the The Andaman & Nicobar Islands, *JBNHS* Vol. 61.

Abdulali, H. (1979). Birds of Great and Car Nicobar with some Notes on Wildlife Conservation in the Island, *JBNHS* Vol. 75 (3).

Anon, (1983). *Hundred Years of Forestry in Andaman (1883-1983).* Forest Department, Andaman and Nicobar Islands.

Anon, (1983). *Red Oil palm Plantation in the Andaman and Nicobar Island - Environmental Assessment,* Department of Environment, Government of India (Mimeo).

Anon, (1985). *Environmental Problems of the Marine and Coastal Areas of India: National Report (No. 59).* United Nations Environment Programme.

Anon, (1985). R*eview of Management Practices in the Forests of Andaman and Nicobar Islands,* Forest Department, Port Blair.

Balakrishna, N.P. (1979). Recent Botanical Studies in Andaman and Nicobar Islands' *BBSI* Vol. 19.

Biswas, S. (1984). Some Notes on the Reptiles of the Andaman and Nicobar Island's, *JBNHS* Vol 81 (2).

Chaturvedi, M.D. (1983). Forests of Nicobar Islands, Ministry of Food and Agriculture.

Chaturvedi, Y. (1980). Mammals of the Andamans and Nicobars, their Zoogeography and Faunal Afinity, RZSI Vol. 77.

Chaudhuri, A.B. (1987). Protected areas of Andaman & Nicobar Islands.

Chaudhuri, A.B. (1991). Plants, Wildlife and Man.

Chengappa, B.S. (1963). Aboriginal Tribes of Andamans, The Indian Forester Vol. 89.

CMFRI - Bull-34 (1983). Mericulture potential of A & N Islands.

Day, F. (1870). Fishes of Andaman & Nicobar Islands.

Das, A.K. (1986) : Estuarine Fauna of Andaman and Nicobar Islands: State of Art Reports, Estuarine Biology, Zoological Survey of India (Mimeo) 27.

Jayal, N.D. (1986) : Andaman and Nicobar Islands - Our Unique Tropical Heritage (Mimeo)

Menon, P.M.G. (1976): Fisheries in the Andaman's, Yojana Vo1. 20 (13)

Osmaston, B.B. (1907) : A visit to South Sentinel Island, *JBNHS* Vol. 17(1,2)

Parkinson, C.E. (1923) : A Forest Flora of the Andaman Islands.

Das, A.K. & Dev Roy, M.K. (1989) : Mangrove Fauna of Andaman and Nicobar (ZSI. Fana. Con. Sens. 4).

Index